BUS

An Introduction to the Technology of Pottery

66(
Ra
AN INTRODUCTION TO THE
TECHNOLOGY OF POTTERY

SECOND EDITION

THE INSTITUTE OF CERAMICS TEXTBOOK SERIES

Editor

R. FREER, University of Manchester/UMIST

OTHER TITLES IN THE SERIES

Published by Pergamon Press on behalf of The Institute of Ceramics

G.C. BYE
Portland Cement

R.W. FORD
Ceramics Drying

THE INSTITUTE OF CERAMICS
Health and Safety in Ceramics, 2nd Edition

K.A. MASKALL & D. WHITE
Vitreous Enamelling

W. RYAN & C. RADFORD
Whitewares: Production Testing and Quality Control

J.R. TAYLOR & A.C. BULL
Ceramics Glaze Technology

W.E. WORRALL
Ceramic Raw Materials, 2nd Edition

Published by The Institute of Ceramics

D.J. CLINTON
A Guide to the Polishing and Etching of Technical and Engineering Ceramics

Pergamon Title of Related Interest

W. RYAN
Properties of Ceramic Raw Materials, 2nd Edition

Pergamon Related Journals *(Free sample copies gladly sent on request)*

International Journal of Solids and Structures

Journal of Physics and Chemistry of Solids

Journal of the Mechanics and Physics of Solids

An Introduction to the Technology of Pottery

SECOND EDITION

Paul Rado, F.I.Ceram.
Formerly Research Manager,
The Worcester Royal Porcelain Company Ltd.,
Worcester, U.K.

Published on behalf of

THE INSTITUTE OF CERAMICS

by

PERGAMON PRESS

OXFORD · NEW YORK · BEIJING · FRANKFURT
SÃO PAULO · SYDNEY · TOKYO · TORONTO

U.K.	Pergamon Press, Headington Hill Hall, Oxford OX3 0BW, England
U.S.A.	Pergamon Press, Maxwell House, Fairview Park, Elmsford, New York 10523, U.S.A.
PEOPLE'S REPUBLIC OF CHINA	Pergamon Press, Room 4037, Qianmen Hotel, Beijing, People's Republic of China
FEDERAL REPUBLIC OF GERMANY	Pergamon Press, Hammerweg 6, D-6242 Kronberg, Federal Republic of Germany
BRAZIL	Pergamon Editora, Rua Eça de Queiros, 346, CEP 04011, Paraiso, São Paulo, Brazil
AUSTRALIA	Pergamon Press Australia, P.O. Box 544, Potts Point, N.S.W. 2011, Australia
JAPAN	Pergamon Press, 8th Floor, Matsuoka Central Building, 1-7-1 Nishishinjuku, Shinjuku-ku, Tokyo 160, Japan
CANADA	Pergamon Press Canada, Suite No. 271, 253 College Street, Toronto, Ontario, Canada M5T 1R5

First edition 1969

Second edition 1988

Library of Congress Cataloging in Publication Data
Rado, Paul.
An introduction to the technology of pottery/Paul Rado.—
2nd ed. (Institute of Ceramics textbook series)
Includes index.
1. Pottery. I. Institute of Ceramics (Great Britain)
II. Title. III. Series.
TP807.R27 1988 666'.3—dc19 87–21299 CIP

British Library Cataloguing in Publication Data
Rado, Paul
An introduction to the technology of
pottery.—2nd ed.
1. Ceramics
I. Title II. Institute of Ceramics
666   TP807

ISBN 0–08–034932–3 (Hardcover)
ISBN 0–08–034930–7 (Flexicover)

Printed in Great Britain by A. Wheaton & Co. Ltd., Exeter

Preface to the Second Edition

Tremendous strides have been made in the manufacture of pottery since the first edition of this textbook was written nearly twenty years ago. The development of new processes, coupled with the application of computers, has transformed into reality what, twenty years ago, was only a conceptualized dream, viz. the continuous conveyor belt production line.

Compared with the first edition of the book, the scope of the second edition has been widened to embrace innovations concerning more efficient shaping methods, energy-saving kiln designs, etc. Moreover, the chapter on the properties of pottery has been brought up to date and elaborated, especially regarding the advances achieved in making pottery decoration dishwasher-proof and in reducing the amounts of lead released from pottery glazes and colours and passed into food to much less than those contained in the food itself (a topic that hardly existed when the first edition was published).

I would like to record my gratitude to Dr. W. E. Worrall of Leeds University for suggesting an updated version of the book in the first place; without his encouragement this revised edition would not have come about. I am grateful to Dr. J. H. Sharp of Sheffield University and especially to Dr. Robert Freer of Manchester University for their editorial assistance. I am greatly indebted to British Ceramic Research Ltd. for having given me permission to use the facilities of that organization's Information Department, my special gratitude being due to Dr. P. Engel and to Mrs. A. Pace for their invaluable help. I also thank my former colleagues at the Worcester Royal Porcelain Division of Royal Worcester Spode for having allowed me to use that company's technical library. Lastly, I wish to express my appreciation to the referee for his constructive criticism and suggestions.

P. RADO

Preface to the First Edition

The term "pottery" means different things to different people. In the present context it embraces domestic ceramic ware, *objets d'art* as well as tableware, ranging from crude pottery to fine bone china.

There have been enormous technical developments in pottery during the last few years and the industry is being more and more mechanized and automated. It is not the purpose of this book to describe engineering aspects in detail—machines will continue to be changed and improved. In an effort to afford the reader an understanding of pottery as a technology, traditional methods of fabrication—particularly where the structure and properties of clay, the basic pottery material, become more immediately apparent—have been dealt with in greater detail than their present application seems to warrant. It is hoped that the resulting blend of information will be of interest to the keen general reader as well as the student professional.

Dr. Peter Murray (who was entrusted with selecting an author for the present volume in the series on ceramics when he was still Assistant Director of A.E.R.E., Harwell) thought a book on pottery should be written by a member of an industrial pottery, preferably one which combines a tradition of fine craftsmanship with a progressive technical outlook. He therefore approached the chairman of the Worcester Royal Porcelain Company, Mr. Joseph F. Gimson, C.B.E. (now retired), whom he knew as a pioneer in the industry. I feel honoured having been asked by Mr. Gimson to write this book and would like to record my gratitude to him for his encouragement.

I am indebted to Mr. A. T. Wright, Works Director of the Worcester Royal Porcelain Company, for his interest and for allowing me to use the Company's library and other facilities. I am grateful to the Trustees of the Dyson-Perrins Museum for permission to reproduce objects from their collection. It is with special gratitude that I acknowledge the advice and help I received from Dr. F. Bäuml, Staatliche Porzellan-Manufaktur Nymphenburg, who granted me permission to include a Bustelli figure. My thanks are due to Dr. Peter Murray for initial discussions. Above all I must thank Dr. G. Arthur for reading and re-reading the manuscript, for his constructive criticism, suggestions, advice, and discussions.

P. RADO

Contents

Chapter 7.   Glazing   124

Chapter 8.   Decoration   150

Chapter 9.   Types of Pottery

Chapter 10.   The Properties of Pottery

List of Figures

List of Tables

1

The Development of Pottery

1. Introduction

Pottery is the perfect combination of what the ancient Greeks regarded as the *four elements* of which the world was made. Pottery is made of *earth*, shaped with *water*, dried in *air*, and made durable by *fire*. Its manufacture comprises the same steps as baking, viz. grinding, mixing with water, kneading, shaping, drying, and firing. In the Neolithic age, ovens for baking bread served also for the firing of pottery. In baking, the product of the earth is used. With pottery the earth itself serves as raw material.

The essential point in the making of pottery is that, like baking, the product is shaped in the cold state and then exposed to heat. As pottery developed it was embellished by patterns, colouring, and glazing. Most pottery is nowadays covered with a glaze. This is applied as a finely ground suspension of the mixed glaze materials to the fired pottery, again in the cold state. In a second firing a thin layer of glass, the actual glaze, is formed on the pottery. Decoration is applied by colouring with metal oxides, usually after glazing. A further firing permanently fixes the colours to the glaze.

2. Man and Clay

The most abundant, ubiquitous, and accessible material on the earth's crust is clay. Early man was in constant touch with it. Having observed that, after heavy rains, clay left his footprints, he found he could shape it with his hands. He thus discovered that clay was plastic. Plasticity, the most important property of clay, made early man realize the potentialities of this material which was the key to the foundation of pottery. He found that objects made from clay retained their shape and when left in the sun became dry and firm.

Early man may have thrown a pot into a fire in order to destroy it; or he may have dropped it into a fire accidentally; or he may have had the idea of baking it like bread. It must have been a thrilling experience for him to find that the pot which was exposed to heat had become much stronger and emitted a pleasing sound when struck. He noticed that his pot had shrunk. Subconsciously he became aware of the phenomena due to firing (sintering, chemical reactions, formation of a glassy phase) and in the same way had already realized the characteristics of plasticity.

1

3. The Archaeological Significance of Pottery

Pottery was the first synthetic material to be discovered by man: an artificial stone produced by firing clay shapes to a temperature sufficiently high to change the physical and chemical properties of the original clay into a new substance with many of the characteristics of stone. It is the stone-like property of pottery which makes it invaluable to the archaeologist, for even if a jar is broken into pieces the fragments are imperishable.

Pottery first appeared about 15 000–10 000 B.C. with the dawn of the Neolithic age. Nomads knew pottery but did not use it as it was too fragile. Pottery is thus a characteristic—almost a symbol—of the settled life. Its appearance and development marks an important stage in the progress of man.

Archaeologists depend largely on pottery types to distinguish various cultures and to establish their chronological order. Pottery can justly be called the calendar of prehistory:

> So thoroughly have archaeologists catalogued its changing styles that the sherds culled from ruins may date a city's rise, zenith and fall within half a century. [Wright, 1960.]
> Bits of pottery are especially welcome as a dating tool in the ancient Mediterranean. Styles of pottery sometimes changed nearly as drastically and as often as styles of clothing in our times; thus stylistic comparison of potsherds from one site with pottery found and dated elsewhere in the Mediterranean enables the experts to place their dates within a century. [Throckmorton, 1962.]

4. Development of Pottery through the Ages

It has generally been assumed that pottery was preceded by basket making. Prehistoric man, or rather woman, strengthened basketwork bowls by smearing the outside with clay. The idea of pottery may have started when such a basket was accidentally burnt. Fired clay pieces with basketwork imprints dating from 15 000 to 10 000 B.C. have been found in Gambles Cave in Kenya. The marks left by the wicker basket acted as a rough decoration and later those markings were deliberately copied on wheel-thrown pots. Some authorities dispute the theory of basket making antedating pottery on grounds of insufficient evidence; they claim pottery as the forerunner of basket making.

The other most common freehand method of modelling pottery was coiling (the building up of vessels by long coils).

Casting involving liquid clay was used in ancient Palestine but this method has been used in Europe only since about A.D. 1730.

The introduction of the potter's wheel, probably man's oldest machine, marked the beginning of the mechanization of pottery. The earliest known use of a potter's wheel was in the Mesopotamian town of Worka (5000 B.C.). Potters also used wheels during the Indus Valley civilization (2500–1500 B.C.), as did the Mayan people for making ceramic toys for their children. By

Old Testament times (2000 B.C.) the potter's wheel was in common use in southern Palestine.

Much of ancient pottery was unglazed, although glazes were known to the Egyptians as far back as 12 000 B.C. There is practically no information about the firing of ancient pottery. It was the most mysterious and the most skilful part of the potter's craft and, as such, kept a trade secret. We have, however, an idea of the difficulty in firing and the wastage resulting from failures, i.e. cracked or misshaped ware, as one of the gates of Jerusalem was called the "Gate of Potsherds" (i.e. broken pots). The Ancient Greeks sought the aid of the gods and medieval potters offered prayers before firing.

Not all pottery was fired. The Indians in North America prevented their dishes and platters, not used for cooking, from cracking by applying grease until the clay was saturated. Clay tablets bearing inscriptions were usually only fired accidentally as a result of an outbreak of fire, as were building bricks for houses.

Concurrently with improvements in the methods of making, refinements were also being made in the raw materials used.

Most clays are contaminated, not only with iron oxide but with sand and feldspathic and calcareous minerals. These reduce plasticity and with some very fine clays this is actually an advantage as otherwise they would be too plastic; pieces made from these over-plastic clays would crack during drying and firing. Where man found such fine clays without the "contaminating" minerals he learned to add non-plastic materials. Those nearest at hand were not only minerals such as sand, limestone, etc., but also straw. Colorado

TABLE 1.1.   *Characteristics of the Most Important Types of Pottery.*
A. *Porous Pottery*

| Type | Raw materials (%) | Firing temperature (°C) | | Colour and special characteristics |
		First fire (biscuit)	Glazing fire	
Common pottery	Clay (impure)	900	1000–1100	Brown-red Sometimes unglazed
Majolica	Clay (impure)+sand and fluxes	900	1000–1100	Brown but covered by white opaque glaze
Earthenware	Ball clay 50 China clay Feldspathic minerals 5–20 Flint or other silica minerals 30–45	1050–1150	950–1050	Off-white

B. *Dense Pottery*

| Type | Raw materials (%) | Firing temperature (°C) | | Colour and special characteristics |
		First fire (biscuit)	Glazing fire	
Stoneware	Clay (naturally fluxed) or clay + fluxes and silica materials	1100–1300	1000–1100	Grey, buff, etc. Sometimes no glazing fire, glaze effect obtained by introducing common salt at end of first fire
Vitreous china	Ball clay ⎫ China clay ⎭ 50 Feldspathic minerals 10–20 Quartz 35–45	1100–1250	1000–1100	Off white–white Slightly translucent
Soft porcelain[a]	China clay 30–40 Feldspar 30–40 Quartz 25–35	900–1000	1250–1350	Off white–white Translucent
Hard porcelain	China clay 50 Feldspar 15–25 Quartz 15–35	900–1000	1400	Bluish white Translucent
Bone china	China clay 25 Cornish stone 25 Bone ash 50	1250	1100	Pure white Translucent

[a] There are several types of soft porcelain (Chapter 9, p.181).

Indians as well as other potters were able to prevent cracking by incorporating finely crushed bits of fired pottery.

These auxiliary minerals also affect the behaviour during firing, most of them acting as fluxes. A very powerful flux is iron oxide, invariably present in the clays nearest the surface of the earth, with which man was in almost direct contact. The iron oxide is also responsible for the firing colour, normally various shades of browns and reds.

In his search for refinement and perfection the potter strove to make his pots white and he first succeeded in doing so by covering them with an opacified glaze: the result was *majolica*. By becoming more selective in his choice of clays, by incorporating greater amounts of white burning non-clay materials such as flint, and by increasing the firing temperature, he produced *earthenware*. On the other hand, by using clays rich in fluxing contaminants and firing sufficiently high he arrived at a non-porous product, *stoneware*. When, however, his clay as well as the other minerals were very low in iron oxide and when he fired to a very high temperature, the resulting material was not only non-porous but also white and translucent. This was given the name "porcelain", the potter's crowning achievement.

Approximate characteristics of the most important types of pottery are shown in Table 1.1

Further Reading

History of ceramic technology: Litzow (1982).
Archaeology and pottery: Chatterjee (1960); Kelso and Thorley (1941–3), (1950); Matson (1965); Shepard (1956).
Method of dating: Aitken (1964); Weaver (1967); Woolley (1937).
Art historical: **Savage (1959)**, (1963); Charles (1964).
General ceramics: Dale and German (1964); Kingery (1960); Norton (1952); Rhodes (1973); Rosenthal (1949); Salmang (1961); Seger (1902); **Singer and Singer (1963)**.

2

Raw Materials

1. Plastic Materials: Clays

Pottery would not have been developed without a plastic raw material. Plasticity has been defined as:

> the property that enables a material to be changed in shape without rupturing by the application of an external force, and to retain that shape when the force is removed or reduced below a certain value. [Moore, 1965.]

Clay has this characteristic. It flows easily at high but not at low stresses. It deforms easily when "worked" but does not deform once it has been made into a shape—its yield value is high.

The plasticity of clay is mainly derived from its content of colloidal particles. This, in turn, largely depends on the type of clay mineral.

1.1. *The Clay Minerals*

1.1.1. *Formation, Constitution, and Properties of Clay Minerals*

The most important clay mineral is *kaolinite* ($Al_2O_3.2SiO_2.2H_2O$).* The name is derived from "Kao-Lin", a hill in North China where a very pure, white-firing clay was first discovered. There are deposits of similar clays in other parts of the world, although they were discovered much later. In Britain this type of clay is known as "china clay"; in America and elsewhere as "kaolin".

Kaolinite is the product of the breakdown of the mineral feldspar or of similar metastable alumino-silicate minerals that were formed under high temperature conditions as components of various kinds of igneous and metamorphic rock. The breakdown occurred in the presence of water at low temperatures when these minerals were no longer in equilibrium with their environment.

In some deposits of china clay, such as those in Cornwall, the mineralogical

---

* Chemical formulae in this textbook are generally expressed in "oxide" form rather than the structurally more correct form (e.g. $Al_2Si_2O_5(OH)_4$ for kaolinite) because this notation will help the reader to understand the "ceramic" role of the various minerals.

evidence shows that the potash feldspar first decomposes to secondary mica and then to kaolinite. The reactions are as follows:

$$3(K_2O.Al_2O_3.6SiO_2) + 2H_2O \rightarrow$$

potash feldspar

$$\rightarrow K_2O.3Al_2O_3.6SiO_2.2H_2O + 2K_2O + 12SiO_2$$

mica

$$+4H_2O$$
$$\rightarrow 3(Al_2O_3.2SiO_2.2H_2O) + K_2O.$$

However, direct conversion of feldspar into kaolinite is also possible by leaching out the alkalis and expelling $4SiO_2$. Mayer and Kästner (1930) claim to have converted a granitic feldspar to kaolinite within a year simply by burying it in garden soil.

The structure of kaolinite (the most perfect of the clay minerals) can be considered as being made up of two layers. One layer consists of silicon and oxygen ions in which each silicon ion is at the centre of a tetrahedron formed by four oxygen ions (*silica layer*, Fig. 2.1). The other layer consists of aluminium and hydroxyl ions (*gibbsite layer*, Fig. 2.2), the hydroxyls forming the corners of an octahedron. When the two layers are combined a layer of kaolinite is formed (Fig. 2.3). Each aluminium ion is now surrounded by six anions, some oxygen and some hydroxyl ions.

The kaolinite lattice is rarely as perfect as shown in Fig. 2.3. Disorder often occurs in the stacking of the layers and this influences the ceramic properties. Thus *halloysite*, which belongs to the group of kaolinitic clay

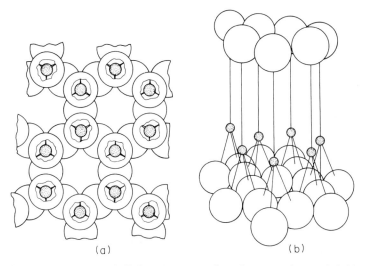

(a)   (b)

FIG. 2.1.   Silica layer unit Si_2O_5: (a) cutaway view, (b) perspective, exploded in c direction. (After Hauth (1951), p. 137.)

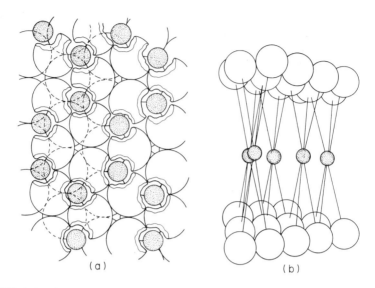

FIG. 2.2.   Gibbsite structure Al(OH)$_3$; (a) cutaway view, (b) perspective, exploded
in c direction. (After Hauth (1951), p. 138.)

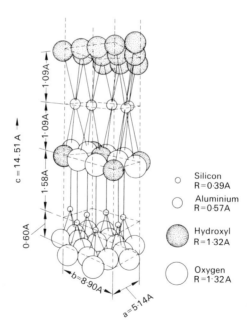

FIG. 2.3.   Kaolinite lattice: (OH)$_4$Al$_2$ Si$_2$O$_5$, perspective, exploded in c direction.
(After Hauth (1951), p. 140).

minerals, has the same structural formula as ideal kaolinite $(Al_2(OH_4)Si_2O_5)$ but its layers are stacked up on top of each other in a different way. In one form of halloysite water molecules exist between the layers; this type is referred to as "hydrated halloysite" $(Al_2(OH_4)Si_2O_5.2H_2O)$.

The other main group of clay minerals of interest in pottery are the *montmorillonites*. They are composed of the same basic constituents as kaolinite but the ratio silica–alumina is twice as great, as shown in the structural formula $(OH)_2Al_2(Si_2O_5)_2$ (or $Al_2O_3.4SiO_2.H_2O$). In montmoril-lonite one layer of aluminium hydroxide is condensed with two silica layers. This is shown in Fig 2.4 giving the structure of the idealized or parent mineral pyrophillite corresponding exactly to the above formula. In other forms of montmorillonite the aluminium hydroxide layer is replaced by magnesium hydrate (brucite) or ferrous hydroxide.

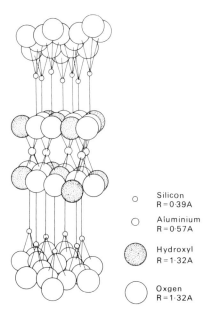

○ Silicon
R = 0·39A

○ Aluminium
R = 0·57A

⬤ Hydroxyl
R = 1·32A

◯ Oxgen
R = 1·32A

FIG. 2.4. Montmorillonite lattice: Pyrophillite structure $(OH)_2Al_2(Si_2O_5)_2$, per-spective, exploded in *c* direction. (After Hauth (1951), p. 141.)

Kaolinite consists of small hexagonal plate-like crystals about 0.1–2 μ in size, only visible under the electron microscope. These crystals are often compressed to concertina-like stacks. In hydrated halloysite the inter-layer water reduces the bonds between layers to such an extent that the layers curl up into small tubes. The crystals of montmorillonite are extremely small, extending from 0.01–2 μ.

The crystals of clay minerals thus lie within the colloidal range (0.001–1 μ) and therefore have colloidal properties, e.g. when dispersed in water or other liquid they do not obey ordinary sedimentation laws but are subject to *Brownian* movement. The smaller the particle size the greater the disorder. The purest clay mineral, viz. non-disordered kaolinite as present in china clay, has the largest particle size and montmorillonite, the most disordered, the smallest particle size of all clay minerals.

In clay minerals, especially the montmorillonites, a certain amount of ionic substitution occurs. For example some Si^{4+} ions in the silicate layers can be replaced by Al^{3+} ions as these two ions are of similar size. Substitution of Al^{3+} ions in the aluminium hydroxide layer by Mg^{2+} ions also takes place. As these substitutions involve the replacement of one cation by another of lower valency the clay structure becomes negatively charged. This charge is compensated by positive ions, e.g. K^+, Na^+, Mg^{2+}, Ca^{2+}, and H^+, absorbed at the edges of the clay minerals or between the layers. These ions are loosely held in the structure and as they can readily be replaced, one by another, are known as "exchangeable bases".

This reaction is known as "cation exchange" and can occur in all clay minerals. It increases with increasing disorder and is of considerable practical importance in the liquid shaping (casting) of pottery. Plasticity can also be increased by replacing a flocculating exchangeable cation such as Ca^{2+} for a deflocculating cation, e.g. Na^+ *or* K^+.

Green strength (the mechanical strength of dry clay) is another ceramic property greatly affected by cation exchange. Contrary to plasticity, green strength is normally decreased when a flocculating cation (Ca^{2+} or Mg^{2+}) replaces a deflocculating one (Na^+ or K^+) because the packing density is thereby decreased. However, the other factors influencing plasticity affect green strength in the same way. Thus greater fineness improves green strength as well as plasticity.

1.1.2. *The Action of Heat on Clay Minerals*

The lattice of kaolinite starts to break up at 420°C when the water of constitution begins to be expelled. Most of it is removed at 550°C, leaving a non-crystalline material known as meta-kaolin. This reaction is endothermic. On further heating exothermic reactions occur around 950°C and 1100°C.

Brindley and Nakahira (1959) by applying single crystal X-ray techniques have shown that the first exothermic reaction corresponded to the formation of a cubic spinel-type phase formed from meta-kaolin by a process of orderly recrystallization. This spine,* probably $2Al_2O_3.3SiO_2$, decomposes and mullite is formed. Crystallization of mullite ($3Al_2O_3.2SiO_2$) is complete at about 1400°C.

On heating, cristobalite starts to form at 1050°C. Silica is progressively discarded as the spinel–mullite reaction process progresses. This implies a far

*The structure of spinel is shown in Fig 8.1, p. 151.

more logical chain of reactions than the views previously held. The dominant feature of the entire reaction series is structural continuity which persists from kaolinite through meta-kaolin to the spinel-type phase.

Brindley sums up the thermal effects accompanying the reactions as follows:

ca. 500°C	Endothermic reaction	Dehydration of kaolinite and formation of meta-kaolin ($2Al_2O_3.4SiO_2$).
ca. 925°C	Exothermic reaction	Meta-kaolin layers condense to form spinel-type phase of approximate composition $2Al_2O_3.3SiO_2$ with discard of silica (about 1 in $4SiO_2$); a "sharp" transformation.
ca. 1050–1100°C	Exothermic reaction	Spinel-type structure transforms to a mullite phase, precise composition not certain, with further discard of silica, appearing as cristobalite.
1200–1400°C		Continued development of cristobalite and mullite, the latter with lattice parameters consistent with the composition $3Al_2O_3.2SiO_2$.

Further work showed that the dehydroxylation was a diffusion controlled process. The dehydroxylation of kaolinite begins on the surface and uniformly progresses inward up to 60 per cent dehydroxylation when the meta-kaolin formed closes the interlamellar channels, leaving isolated patches of kaolinite from which the water escapes only with difficulty (Brindley and others, 1967). This assumption explains why higher temperatures are required to remove all the OH groups from the kaolinite structure (Criado and others, 1984).

A new model for meta-kaolin has been proposed, consisting of anhydrous regions of distorted Al–O tetrahedra containing randomly distributed isolated residual hydroxyls, associated with Al–O configurations of regular octahedral and tetrahedral symetry. (Such a structure accounts for the lack of a well-defined X-ray pattern and the persistence of about 10 per cent residual hydroxyls in meta-kaolin (Mackenzie and others, 1985).)

Brett and others (1970) have suggested in their comprehensive review on thermal decomposition of hydrous layer silicates that the rate-determining step in the formation of mullite from kaolinite was nucleation controlled.

With *montmorillonite* the break-up of the lattice occurs at 700°C. The remaining constituents, silica, alumina, and magnesia, first present in an amorphous mass, form mullite ($3Al_2O_3.2SiO_2$), cristobalite (SiO_2), cordierite ($2MgO.2Al_2O_3.5SiO_2$), and spinel ($MgO.Al_2O_3$) on being heated to 1200°C.

The other clay minerals react in a similar way when heated. Natural clays rarely consist of only one clay mineral. Often they contain *mica*, usually in the form of *muscovite* ($K_2O.3Al_2O_3.6SiO_2.2H_2O$) and sometimes *illites*, also referred to as *sericites* or *hydrous micas*, being structurally similar to mica but containing less potash and more combined water. Mica loses its water of combination at 800°C. The difference in temperature at which kaolinite and mica give off their combined water has been used as a means of identifying the two minerals. Between 1100°C and 1200°C mica forms glass and mullite.

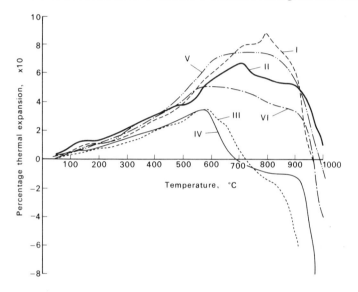

FIG. 2.5.   Length changes of some English clays on firing up to 1000°C. I, Siliceous ball clay; II, Black ball clay; III, South Devon ball clay; IV, China clay; V, Red tea pot clay; VI, Fireclay. (After Dale and Francis (1943), p. 42.)

Foreign ions, no matter whether present in the lattice of the clay mineral or as discrete mineral impurities in the clay, increase the rate of reaction. They act as "fluxes" or as mineralizers, as, for instance, titania.

The action of heat on clay minerals brings about changes in length. These are shown for various types of clay in Fig. 2.5.

1.2. *China Clays*

1.2.1. *Origin and Characteristics of China Clays*

China clays are the purest clays. Most of them consist of kaolinite and do not contain any other clay mineral, although in a few china clays the kaolinite may be of the disordered type.

The kaolinization process (the transformation from feldspar to kaolinite), mentioned under "Formation, Constitution, and Properties of Clay Minerals" (p. 6), is not the only theory advanced for the conversion of feldspar to kaolin. It is true for the Cornish deposits. French deposits of china clay are believed to have been formed by surface weathering, i.e. downward percolating water containing carbon dioxide. (Mellor (1935) described the very mild insidious action of water followed by the decay and weathering of feldspar, as "nature's favoured process of making clay".)

China clays are *primary* or *residual* clays. They are still at the site of their formation. This is one of their characteristics, although the white-burning clays of South Carolina, Georgia, and Florida are referred to as "sedimentary kaolins". "Sedimentary" usually implies secondary deposits which have been washed away from the site of their formation. Most clays other than china clays are sedimentary.

China clay was not discovered by Western man until the eighteenth century A.D. Being almost free from discolouring oxides it fires to a white colour. This property distinguishes it from other clays and made the manufacture of porcelain and china possible.

From a geological point of view china clays are very young clays and kaolinization is still taking place. The yield of kaolinite is relatively small. (In Cornwall the amount of china clay is only about 10–20 per cent of the parent rock.) For this reason the "raw" china clay has to be refined and the china clay proper separated from other minerals.

1.2.2. *Deposits, Winning, and Purification of China Clay*

China clays (or kaolins) are found in many parts of the earth. In Europe the deposits near Zettlitz in Czechoslovakia used to be regarded as the most plastic, if not the purest. Relatively plastic kaolins are also found in Spain and Portugal, as well as in Saint-Yrieix en Limousin in France. The German kaolin deposits in Saxony and Bavaria are less plastic and less pure. Of non-European china clays those in Georgia, U.S.A., are the best known. There are considerable deposits in China, India, and North Africa. Perhaps the whitest-burning kaolins are the china clays of Cornwall, their iron and titanium oxide content being very low. Typical chemical analyses—in descending alumina contents—are given in Table 2.1.

Within the last few years china clays consisting of halloysite have come on the market. The halloysitic kaolin of Burela, Spain, is reported to be one of the constituents of almost all high-quality porcelains manufactured in the Federal Republic of Germany (Schüller, 1976)—as is English (kaolinitic) china clay. The halloysite mined in New Zealand is claimed to contain less impurities, especially iron oxide and titania, thus producing whiter and more translucent porcelain than do kaolinitic china clays (Townsend and Luke, 1985). However, the halloysitic materials showed insufficient green strength and plasticity and had to be partially replaced by special plastic china clays (Wood, 1985).

Methods of winning and beneficiation vary according to the depth and the extent of the deposit, as well as the depth of the overburden. If this is only a few feet thick it is mechanically stripped and open-cast mining is applied. If it

TABLE 2.1. *Chemical Analyses of China Clays*

Grade:	Kaolinite	Standard Porcelain	CC 31	Pleyber S	Diamond
Source:		ECC International Ltd. 1984	Watts, Blake, Bearne & Co. plc. 1985	ECC International Ltd. 1984	Goonvean & Rostowrack China Clay Co. Ltd. 1984
Location:		Cornwall, England	Cornwall, England	Brittany, France	Cornwall, England
SiO_2	46.60	48	47.6	47.5	48.30
TiO_2	—	0.02	0.1	0.3	0.05
Al_2O_3	39.50	37	36.7	37.5	36.40
Fe_2O_3	—	0.65	0.9	0.65	0.58
CaO	—	0.07	0.1	0.04	0.20
MgO	—	0.3	0.3	0.22	0.38
K_2O	—	1.6	2.3	1.5	2.00
Na_2O	—	0.1	0.1	0.1	0.25
LOI	13.90	12.5	12.0	12.4	11.84
Total	100.00	100.24	100.1	100.21	100.00

is of great depth, as with Zettlitz kaolin, shafts are sunk from which alleys and galleries radiate in a similar way as in coal mining.

The mined lumps of crude kaolin, containing about 75 per cent coarse quartz sand and fine mica, are crushed and fed into rotating drums where the first separation takes place, the coarse fractions, i.e. quartz and mica, settling out. This process of separation by gravity is greatly accelerated by centrifugal forces as in a hydrocyclone; this consists of a tubular centrifuge with a stationary barrel in which the kaolin suspension rotates; the rotating motion is brought about by tangential feed of the suspension under pressure, as shown in Fig. 2.6 (Trawinski, 1979). A general flow sheet is given in Fig. 2.7 (Trawinski, 1979).

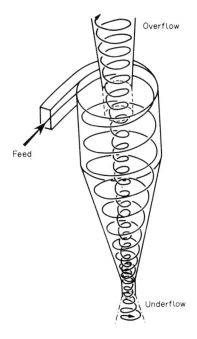

FIG. 2.6.   Hydrocyclone flow diagram. (After Trawinski (1979), p. 4.) Reproduced by permission of Verlag Schmid GmbH.

With the enormous china clay deposits in Cornwall a unique yet simple method is used (Fig. 2.8). The overburden is removed mechanically and once a steep-sided pit has been formed powerful water-jets are directed towards the pit faces. The slurry produced runs into reservoirs at the bottom of the pit in which the coarse extraneous minerals settle out. They are periodically removed, forming conical white dumps characteristic of the Cornish landscape. The clay slurry is pumped by centrifugal pumps to the surface for further refining. It is passed through long channels where the fine sand and mica flakes settle out. Purification is aided and accelerated by centrifuges and hydrocyclones. The final clay suspension, which only contains about 2 per cent solids, is run into settling tanks where surplus water is drawn off. When the suspension contains about 10 per cent solids it is filter-pressed and dried.

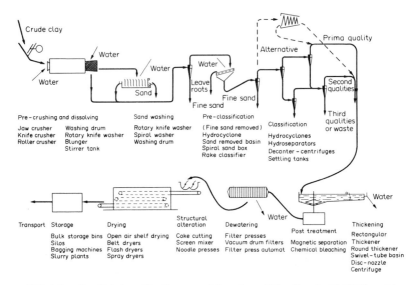

FIG. 2.7.   Beneficiation of kaolin, general flow sheet. (After Trawinski (1979), p. 1.) Reproduced by permission of Verlag Schmid GmbH.

A great deal of research has been carried out on Cornish china clays (a) to improve the properties, such as plasticity, strength in the unfired state, etc., and (b) to make them more consistent (Clark, 1964). Individual pits vary and in an effort to reduce variation, clays from various pits are blended. This is now done entirely by computer (see Fig. 2.9).

The electron microscope microprobe analyser (EMMA) has made it possible to analyse single crystals, and it has thus been established that kaolinite particles have lower iron contents than mica particles (Noble and others, 1979). The latter are larger than kaolinite grains and are, therefore, eliminated by simple sedimentation in the normal refining process. In a further effort to improve the whiteness of china clays a number of methods of beneficiation have been developed, the most effective being *high-intensity magnetic separation*. Titania, which in itself is white but enhances the staining power of iron oxide, has been successfully removed by *froth flotation*, a process in which particles are made hydrophobic (described in greater detail under sections 2.1.2 and 2.2.4 of this chapter). The most efficient method of eliminating titania is perhaps *selective (differential) flocculation*, discovered by accident when starch was used in a clay suspension and separation occurred.

Quality is maintained by the use of on-stream chemical analysers (Wood, 1985).

FIG. 2.8. Overview of a china clay pit in Cornwall. By courtesy of English China Clays Group.

1.2.3. *The Use of China Clays in Pottery*

China clays form the main ingredient of porcelains because they are the only clays whose iron contents are sufficiently low to produce the white colour and translucency in the fired product. They are also the only clay material in bone china and are used in earthenware to improve (lighten) the colour.

1.3. *Sedimentary Clays*

Most clays are secondary which means that they have not remained at their original site of formation. After they were formed from feldspar they were exposed to the further action of water. In this way nature beneficiated them

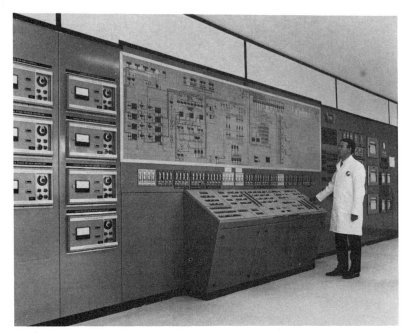

FIG. 2.9.   China clay refining plant control room. By courtesy of English China
Clays Group.

and put them at the disposal of man, ready for use. She made a much better
job than man could have done, by elutriating them extremely finely; this
resulted in far better plasticity and strength in the unfired state than man is
capable of achieving by the refinement of china clays. However, while they
were transported, eventually becoming sedimentary deposits, they became
contaminated by the oxides of the most common metal, iron. Titanium
dioxide, another impurity picked up, although colourless, greatly enhances
the colour of iron oxide in clays. These sedimentary clays are therefore
invariably discoloured when fired. Recalling J. W. Mellor, the great chemist
and ceramist, this was a typical case where: "nature has used dirty beakers
and dirty solvents in the production of minerals and she has not always
preserved her products from contamination after they have been made."

Sedimentary clays do not usually consist of pure kaolinite but of disordered
kaolinite and one or several of the other clay minerals such as halloysite,
montmorillonite, etc.

1.3.1. *Ball Clays*

The name is derived from the original method of winning in which the clay
was cut into balls, each weighing 30–35 lb. Ball clays consist mainly of
kaolinite but have some montmorillonite attached to the edges of the

kaolinite platelets. In the unfired state they are often dark in colour, grey, blue, or even black due to organic impurities. This carbonaceous matter is removed in the early stages of firing. Some ball clays are light coloured or nearly white in the fired state if their iron content is low.

The most important deposits are in Devonshire and Dorset. In the United States ball clay deposits occur in Florida, Tennessee, Kentucky, Alabama, and New Jersey. On the continent of Europe they are found in France, Germany, and Czechoslovakia.

Ball clays vary in chemical composition much more than china clays. The most prevalent impurity is quartz sand, some ball clays containing up to 66 per cent corresponding to 80 per cent total silica. Typical chemical analysese—in ascending silica contents—are given in Table 2.2.

Ball clays are extracted by shaft mining but more often by open-cast mining. The overburden is removed by bulldozers and scrapers or excavators, draglines, and dumpers. The actual digging of the clays is usually done by specially designed pneumatic tools.

Mechanical separation or refining as with china clays is not necessary for ball clays. Weathering is the only treatment, apart from hand-sorting, which some ball clays receive after mining. This applies particularly to very hard clays which are difficult to break down. The clay is merely subjected to exposure to the weather for several months, some Scottish clays up to 3 years. (The effect is similar to that of garden soil which, after having been dug in the autumn, has been exposed to frost and breaks up easily in spring.) This improves the workability, the plasticity, and even the fired colour. Some ball clays contain sulphates which become harmful during firing and these are also successfully removed by weathering. Any chemical treatment of ball clays would be too costly; most of the iron impurities are non-magnetic.

1.3.2. Stoneware Clays

These are a type of ball clay sufficiently high in impurities, mainly alkalis, also alkaline earths, to produce fully dense (vitrified) ware at relatively low temperatures with only little or no additional flux. Stoneware clays are found in many countries and have given rise to local pottery of individual character.

1.3.3. Miscellaneous Types of Clays

Of these the so-called "red" clays are of the greatest interest to the potter. Their distinguishing feature is their high iron content (up to 10 per cent). This acts as a strong flux and, as with some stoneware clays, they do not need additional flux. The firing temperature can be even lower than that of stoneware. Red clays are the most abundant and as they often occur with practically no overburden it is this type of clay man first met.

Marls, natural mixtures of clay and chalk, are friable, their texture being

TABLE 2.2.   *Chemical Analyses of Ball Clays*

Grade:	HVA/R	EWVA	HYMOD PRIMA	HC 5
Source:	Watts, Blake, Bearne & Co. plc. 1985	Watts, Blake, Bearne & Co. plc. 1985	ECC International Ltd. 1981	HC Spinks Clay Company Inc. 1981
Location:	S. Devon	S. Devon	Dorset	Tennessee, USA
SiO_2	61.1	47.6	54	62.72
TiO_2	1.4	0.9	1.4	1.40
Al_2O_3	>6.1	33.1	30	25.96
Fe_2O_3	0.9	1.0	1.2	1.31
CaO	0.2	0.2	0.3	0.08
MgO	0.3	0.3	0.4	0.02
K_2O	2.6	1.7	3.1	0.16
Na_2O	0.3	0.1	0.5	0.17
Loss on ignition	7.1	15.1	8.7	8.02
Total	100.0	100.0	99.6	99.84

quite different from that of other clays. The term "marl" is also applied to clays of similar texture but with very little or no chalk, e.g. Staffordshire marls. Marls are used in the same way as red clays, the CaO content acting as additional flux.

Fireclays, again occurring in many different localities, are seldom used in pottery because they are too hard. Many have to be crushed like rocks before they can be mixed with water. They contain low amounts of fluxing impurities, such as alkalis and alkaline earths, and are therefore more refractory than other sedimentary clays. They are the material for making the refractory containers and other kiln "furniture" in which pottery is fired.

1.3.4. Bentonites

Bentonite, which consists of montmorillonite, the most disordered of the clay materials, is believed to be derived from volcanic ash. Being almost entirely colloidal, like glue, bentonite absorbs water very readily. This causes it to swell up to five times its dry volume. Bentonite can never be used on its own as it is far too plastic. The highest amount introduced is about 5 per cent and that only for special products which contain no other clay material. It is not normally used in pottery, except occasionally in bone china which, owing to its low clay content, lacks plasticity. As little as ½ per cent bentonite added to bone china gives a considerable improvement in plasticity and green strength. More than this would spoil the fired colour because of its extremely high iron content which forms dark brown glasses when fired to bone china temperature.

The best-known deposits of bentonite are in Wyoming, U.S.A.; some are found in Italy.

2. Non-plastic Materials

The function of non-plastic materials is twofold: (i) They are added to clays which are too plastic. As already pointed out in Chapter 1 (p. 3), some clays, having too great a proportion of colloidal particles, tend to crack or distort during drying and firing. Sand or other hard materials, finely divided, added to such clays, reduce the plasticity and thus eliminate these faults. Strange as it may seem, the low plasticity and green strength of china clay are usually enhanced by the addition of non-plastic material because a wider grain-size distribution and hence better packing is provided. (ii) Non-plastic materials make it possible for the desired properties to be obtained at lower firing temperatures. This applies particularly to china clays, which are much purer than other clays. Broadly speaking, the purer the clay the more additional materials are required. With the clays early man first came across there was no need to add non-plastic minerals as nature had already admixed them.

2.1. *Silica*

Silica (SiO_2) is the most abundant oxide found on the earth's crust. It forms silicates with other oxides and the most important pottery raw materials, such as clay, feldspar, Cornish stone, talc, etc., are silicates. On the other hand, silica as such, in one phase or another, forms a very important ingredient in all types of pottery.

2.1.1. *The Phases of Silica*

Silica in the free state usually exists as *quartz* in nature. The other two important phases are *cristobalite* and *tridymite*. As already indicated, when describing the kaolinite crystal (p. 7) each silicon (Si^{4+}) ion is situated in the centre of four oxygen (O^{2-}) ions (Fig. 2.10). If we imagine the centres of the

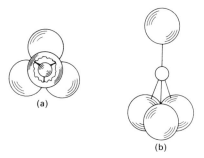

(a)

(b)

FIG. 2.10.   Orthosilicate unit $(SiO_4)^{-4}$: (a) cutaway perspective view, (b) expanded perspective view. (After Hauth (1951), p. 47.) Reproduced by permission of The American Ceramic Society.

four oxygen ions to be joined by straight lines the resulting geometrical figure is a tetrahedron (Fig. 2.11). The tetrahedra are connected to each other in such a way that each O^{2-} ion is effectively joined to two Si^{4+} so that their formula is not (SiO_4) but SiO_2.

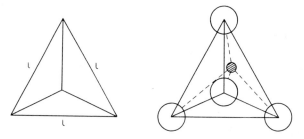

FIG. 2.11.   Silicon–oxygen tetrahedron. (After Worrall (1986), p. 14.) Reproduced by permission of Elsevier Applied Science Publishers Ltd.

In quartz the Si–O–Si bonds joining neighbouring tetrahedra are not straight but bent to form spiral chains (Fig. 2.12). In cristobalite as well as tridymite the silica tetrahedra make up rings containing six silicon atoms and oxygen atoms (Figs 2.13 and 2.14). In both cases one central, common oxygen atom forms the apex of a lower and an upper tetrahedron. In tridymite the three basal oxygen atoms of the upper tetrahedron are situated

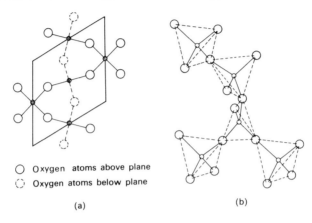

○ Oxygen atoms above plane
◌ Oxygen atoms below plane

(a)   (b)

FIG. 2.12.   Diagrammatic structure of beta-quartz, silicon atoms being represented by the small circles. (After Rigby (1949), p. 18.) Reproduced by permission of The Institute of Ceramics.

exactly above corresponding oxygen atoms in the base of the lower tetrahedron (Fig. 2.13). In cristobalite, on the other hand, one of the tetrahedra is twisted through 60° compared with the other (Fig. 2.14).

The densest packing of the atoms occurs in quartz and it has, therefore, the highest specific gravity of the three forms of silica, viz. 2.65 at 20°C. The corresponding figures are 2.33 for cristobalite and 2.27 for tridymite.

Tridymite rarely occurs in pottery. Cristobalite is present as very small particles, too minute to be seen under the optical microscope; it is usually derived from the kaolinite lattice break-up of china clay, but may also originate, as in earthenware, from quartz (converted) introduced as raw material.

On being heated quartz expands uniformly up to 573°C. Then what is known as the *inversion* occurs: the Si–O–Si bonds become straight; therefore the atoms become less closely packed and a marked expansion occurs (as much as 0.2 per cent, Fig. 2.15) the quartz changing from its "low" to its "high" form. (The terms "low" and "high" are preferable to "alpha" and "beta", sometimes used, because there is no international agreement as to which is alpha and which beta). The other two phases are subject to similar inversions but at lower temperatures: 220–260°C for low to high cristobalite, 117°C for low to high "1" tridymite, and 163° for high "1" to high "2" tridymite. A particularly large expansion occurs at the inversion of cristobalite (Fig. 2.15).

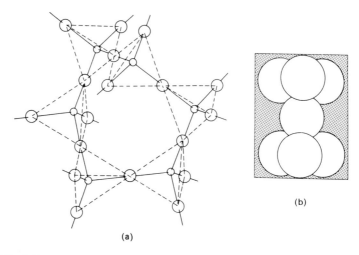

FIG. 2.13. (a) The crystal structure of beta-tridymite, silicon atoms being represented by the smaller circles. (b) The arrangement of the oxygen atoms in tridymite showing hexagonal packing. (After Rigby (1949), p. 19.) Reproduced by permission of The Institute of Ceramics.

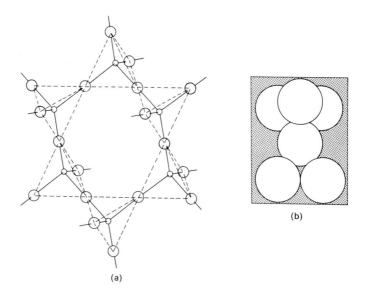

FIG. 2.14. (a) The crystal structure of beta-cristobalite, silicon atoms being represented by the smaller circles. (b) The arrangement of the oxygen atoms in cristobalite showing hexagonal packing. (After Rigby (1949), p. 19.) Reproduced by permission of The Institute of Ceramics.

These inversions are reversible and occur rapidly. However, *conversion* from one phase to another is not easy and takes a long time because this involves separating the tetrahedra and then joining them again differently. Conversions occur well above 600°C (when quartz has ceased to expand and actually slightly contracts).

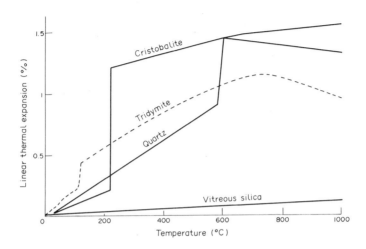

FIG. 2.15. Thermal expansion of the principal forms of silicon (After Worrall (1986), p. 23.) Reproduced by permission of Elsevier Applied Science Publishers Ltd.

If quartz is heated to 1470°C or above, cristobalite is formed. Tridymite can be obtained by heating quartz in the range of 870–1470°C for several days. The tridymite so produced slowly changes to cristobalite if heated above 1470°C. Heating above 1710°C causes conversion to a glass, described as vitreous silica, characterized by an extremely low thermal expansion. The irregular structure of this glass, compared with the crystalline state, is shown in Fig 2.16. (The reactions of conversion can be accelerated by so-called "mineralizers". e.g. lime, which assists the conversion of quartz to cristobalite.)

Quartz is far more common in nature than the other two modifications and is the only phase used (as raw material) in pottery. Cristobalite is present in fired pottery as a result of reaction as already indicated (sometimes due to converted quartz but mostly derived from the break-up of the kaolinite lattice).

2.1.2. *Occurrence and Winning of Silica Materials*

Quartz crystals (rock quartz) occur as individual veins in primary rocks or as grains interspersed between other minerals. Quartz also exists in the sedimentary form as *sand*, as *sandstone* (in which grains of sand are cemented together by fine silica or other substances), or as *quartzite* (which has been formed from sandstone by metamorphosis—the action of heat and pressure).

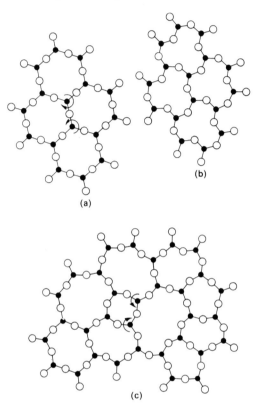

FIG. 2.16.   Crystalline structure (a and b), vitreous structure (c). (After Hauth (1951), p. 203.) Reproduced by permission of The American Ceramic Society.

Flint consists of extremely small crystals of quartz bound together by molecules of water. On heating, the water is evolved at 400°C, causing the large aggregates to become friable, and the quartz changes to cristobalite at 1100°C. Flint often contains lime as an impurity and this promotes conversion. In the United States the term "flint" is generally applied to quartz, finely ground.

The purest and most extensive deposits of crystal quartz in Europe are found in Scandinavia. Deposits of extremely pure crystal quartz are in Brazil. More or less pure sands occur on most continents.

The various types of quartz are mined by conventional quarrying methods. Open cast mining is employed with rock quartz, often with preparatory blasting.

Flint is usually present in the form of pebbles at the sea shore; they are still collected by hand from the Channel coast and classified.

Quartz sands are subjected to a scrubbing and attrition action to remove contaminants; froth flotation is used when a high degree of purity is required, as in white pottery. In this process one mineral species becomes coated with a hydrophobic (water-repellent) reagent, and is thus selectively attached to and carried up to the surface by air bubbles, initiated by frothing agents, rising from the bottom of the flotation cell. The hydrophobic mineral particles are collected in the froth resulting from the bubbles. This froth is continuously skimmed off by paddles (Weiss, 1979).

All types of quartz used in pottery are almost 100 per cent silica, impurities such as ferric oxide and alumina usually amounting to only fractions of 1 per cent. Alkalis in sand very considerably reduce its thermal expansion.

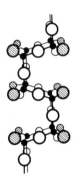

FIG. 2.17.   The structure of the feldspars: part of a feldspar chain. (After Wells, A. F., *Structural Inorganic Chemistry*, Clarendon Press, Oxford.) Reproduced by permission of Oxford University Press.

2.1.3. *The Use of Free Silica in Pottery*

For over 200 years crystalline silica has been regarded as an indispensable beneficial constituent of pottery; however, it has been critically reviewed and been blamed for a number of difficulties in pottery making (Dinsdale, 1963). The indictment is that free crystalline silica:

(a) causes cracking of the ware if fired or cooled too rapidly,
(b) reduces mechanical strength of earthenware and other pottery,
(c) lowers resistance to heat shock in cooking ware,
(d) adversely affects the appearance of glaze,
(e) causes pneumoconiosis, formerly known as silicosis, in pottery workers.

All these faults, except the last one mentioned, arise from the phenomenon of inversion. Despite these criticisms crystalline silica is still being used, no

suitable substitute of equally low price having yet been found. As regards
health hazards the present stringent factory regulations prescribing dust
extractors and other precautionary measures have virtually eliminated
silicosis.

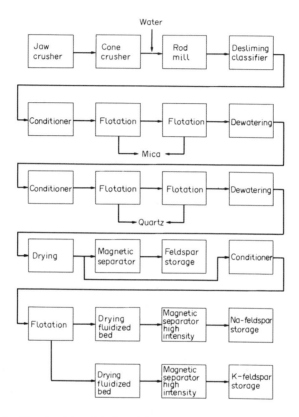

FIG. 2.18.   Flotational beneficiation of feldspar, generalized flow diagram. (After
Williamson (1980), p. 7.) Reproduced by permission of Verlag Schmid GmbH.

The function of silica depends to some extent on the type of pottery into
which it is introduced. In porous pottery, e.g. earthenware, it acts as a filler.
In dense pottery, such as porcelain, it assists the formation of glass (and may
even improve mechanical strength).

On the continent of Europe quartz has been the preferred form of
crystalline silica; in Britain it used to be flint, because it easily converts to
cristobalite, thus giving a high thermal expansion, desirable in earthenware
for good "match" with the glaze (see Chapter 7, section 3, p. 126). Owing to
the difficulties in obtaining flint free from inclusions of chalk, there has been
a tendency to change over to quartz sand. Pottery containing sand has
thermal expansion characteristics different from those of pottery with flint.

This has caused thermal "mismatch" (see Chapter 7, section 3.2, p. 126), as evidenced by "craze" marks on the glaze. This problem has now been solved by grinding the sand less finely.

2.2. Feldspathic Minerals

All feldspathic materials act as "fluxes" in pottery. They form glasses if heated to sufficiently high temperatures, either in themselves or together with the other substances, particularly silica, present in pottery. They are added to decrease the firing temperature and thus to reduce costs.

Feldspathic materials do not have a defined melting point like metals or single oxides. Exposed to heat they start to decompose and soften progressively until they become a viscous mass. This softening and melting is far more gradual in alkali silicates than in alkaline earth silicates. During cooling the liquids (melts) formed are "frozen" and actual glass formation occurs. If more than one alkali or alkaline earth oxide is present the reaction is greatly hastened and melting is completed at much lower temperatures. Typical analyses of feldspathic minerals are given in Table 2.3.

2.2.1. The Feldspars

The most common feldspars used in pottery are:

Orthoclase   (potash feldspar) $K_2O.Al_2O_3.6SiO_2$.
Albite   (soda feldspar) $Na_2O.Al_2O_3.6SiO_2$.

In these alumino-silicates the arrangement of silica and oxygen ions is somewhat different from that of silica. The silica and oxygen atoms are linked so as to form four-membered rings, each containing four oxygen atoms. The rings also contain three silica atoms and one aluminium atom. The four-membered rings are linked with other rings to form chains which are cross-linked with similar chains via Si–O–Si groups, forming a three-dimensional framework (Fig. 2.17). The sodium and potassium atoms are large and situated in cavities within the framework.

Orthoclase (potash feldspar), when heated, begins to decompose at about 1160°C and melting is completed at about 1290°C. Pure orthoclase melts incongruently, breaking down into leucite ($K_2O.Al_2O_3.4SiO_2$) and glass; at 1530°C the leucite crystals are redissolved. Albite (soda feldspar) melts at 1160°C and small amounts of albite in orthoclase appreciably lower the temperature at which melting is completed, a mixture of 65 per cent orthoclase and 35 per cent albite melting at 1070°C.

Anorthite (lime feldspar) ($CaO.Al_2O_3.2SiO_2$), is similar in structure to orthoclase and albite. Its occurrence in nature is rare but it is readily formed as a reaction product in ceramics.

TABLE 2.3. *Chemical Analyses of Feldspathic Minerals*

Type:	Potash Spar	Soda Spar		Nepheline Syenite
Source:	Norfloat A/S 1981	Norfloat A/S 1981	AB Forshammar 1981	Norsk Nefelin 1985
Location:	Norway	Norway	Sweden	North Cape, Norway
SiO_2	65.40	69.20	75.5	57.0
TiO_2	—	—	—	—
Al_2O_3	18.70	18.70	14.70	23.8
Fl_2O_3	0.06	0.11	0.13	0.12
CaO	0.51	1.82	0.20	1.7*
MgO	—	—	0.10	—
K_2O	11.10	2.80	4.25	9.1
Na_2O	3.36	7.20	4.85	7.8
Li_2O	—	—	—	—
LOI	0.29	0.19	0.45	1.1
Total	99.42		100.18	100.62

* This includes 0.30 per cent BaO and 0.30 per cent SrO

Feldspars are igneous minerals occurring in primary rocks and are often mixed with quartz and mica. The most important deposits are in Scandinavia but feldspars are also extensively mined in Canada, Portugal, and South Africa.

2.2.2. Feldspar-bearing Rocks

Nepheline syenite consists of potash feldspar, soda feldspar, as well as nepheline, a rare mineral of the formula $K_2O.3Na_2O.4Al_2O_3.9SiO_2$, and has a structure very similar to that of tridymite. Nepheline syenite was discovered in Canada sometime before the Second World War and has been used there and in the United States for several years. Norwegian deposits opened up a few years ago led to its use in Europe. It replaces feldspar because of its greater fluxing power.

Pegmatite, a mixture of orthoclase and quartz, can be regarded as an impure form of feldspar. It is sometimes used as one single source of feldspar and quartz. However, the proportions of the two minerals vary in some deposits; therefore, feldspar and quartz are separated and supplied individually (see below under section 2.2.4).

Cornish stone, a material confined to England, is a type of pegmatite, but is more varied in its composition, containing both types of alkali feldspars, mica, quartz, and small amounts of tourmaline, fluorspar, and topaz. Because of these fluorine-bearing minerals the use of Cornish stone (or china stone) has declined, fluorine being harmful, not only to pottery itself but also to kiln refractories. Cornish stone has been replaced, to some extent, by synthetic mixtures, representing its chemical composition without fluorine.

Anorthosite, a rock mineral consisting of a solid solution of albite (sodium feldspar) and anorthite (lime feldspar), has shown promise as a suitable raw material for porcelain fired rapidly; it could be a material for the future, bearing in mind the modern trend for short firing cycles (Lyng and Gamlen, 1974).

Mica, discarded in china clay mining, is another possible source of alkali silicate, chemically similar to feldspar. In the past micaceous fractions in china clay have been rejected because of their excessive iron impurities, which would cause dark specks in the fired pottery. However, latest research has shown that coarse micaceous china clay fractions are suitable for replacing about 15–25 per cent of feldspar in earthenware and vitrified hotelware. Their fluxing effect above 1100°C is greater than that of a comparable feldspar/quartz content (Radford and Keating, 1985). This is in agreement with general experience with mica (see Chapter 6, section 1.2, p. 97); Inzigneri and Peco, 1964).

2.2.3. *Lithium Feldspars*

Mention should also be made of the *lithium*-bearing minerals which are relatively new to pottery and are only rarely—and if so sparsely—introduced. They may gain in importance because of their high fluxing power and because when mixed in certain proportions with clays they cause very low, zero, and even negative expansions. This is of great importance where heat shock is encountered, as in pottery cooking vessels. The two lithium feldspars are *spodumene* ($Li_2O.Al_2O_3.4SiO_2$) and *petalite* ($Li_2O.Al_2O_3.8-SiO_2$).

The other lithium minerals used in pottery are *lepidolite* ($K_2O.2Li_2O.-Al_2O_3.8SiO_2.2H_2O$), a mica having a structure similar to the clay mineral montmorillonite (Fig 2.4), and *amblygonite* ($2LiF.Al_2O_3.P_2O_5$), one of the very few phosphates employed in pottery besides bone ash.

2.2.4. *Winning and Beneficiation of Feldspathic Minerals*

Feldspars used to be mined according to conventional quarrying techniques and supplied to the pottery industry in lump form; high-quality grades used to be hand selected. Many of the feldspar mines have been exhausted or have become uneconomical. Fortunately, however, there are vast deposits of pegmatite, especially in Scandinavia. These are being utilized to supply feldspars, produced by froth flotation. A flowsheet is shown in Fig. 2.18 (Williamson, 1980).

The first flotation step is the mica float where all iron-bearing impurities are brought to the surface while feldspar and quartz are depressed. These are passed to a set of conditioning tanks where the reagents, required for their separation, are added. Quartz is then depressed, whereas feldspar is floated off. In some plants potash feldspar and soda feldspar are separated, again by froth flotation. It is claimed that by froth flotation of pegmatite, applying in-line quality control, end products are achieved which are superior to, and more consistent than, the lump material used hitherto (Batchelor, 1973).

2.3. *Alkaline Earth Minerals*

In pottery the alkaline earth minerals mentioned below, like the feldspars, act as "fluxes". Unlike the feldspars which may be present in amounts up to 35 per cent (as much as 70 per cent in "Parian" Chapter 9, p. 184) they are usually only introduced in small quantities, normally less than 10 per cent. Their fluxing action is more powerful than that of feldspars because the proportion of alkaline earth oxides in these minerals, particularly the carbonates, is much higher than the percentage of alkalis in the feldspars.

Although the alkaline earth minerals are fairly refractory they react with silica, invariably present in pottery, to form glasses in a similar way to the feldspars but often at much lower temperatures. Their softening and melting range is much shorter than that of feldspars.

2.3.1. Silicates*

Talc, also known as *steatite* and formerly as *soapstone* ($3MgO.4SiO_2.H_2O$), is composed of a layer of *brucite* ($MgO.2H_2O$) "sandwiched" between two silica-type layers; structurally it thus belongs to the clay mineral group of montmorillonites (Fig. 2.4).

When heated, talc loses its combined water, the last traces being given off at 1050°C; it decomposes into enstatite (MgO,SiO_2) and cristobalite (SiO_2).

Talc consists of secondary rocks formed by interaction of water together with magnesium salts on primary rocks. Chief deposits are in India, China, Egypt, the United States (Wyoming), France, and Germany. It is usually purified by froth flotation.

Unlike quartz and the feldspars, which are brittle and very hard, talc is a soft material (lowest on the Mohs scale) and feels soapy. Although non-plastic, talc can be die-pressed without plasticizer.

Talc was used in pottery in India as early as the Indus Valley civilization (2500–1500 B.C.). Its first recorded use in Europe was in Bristol and Worcester around 1751 (later also at Zurich). Although acting as a flux in pottery if added in small amounts it is much more refractory when used by itself with only a very small amount of clay and an auxiliary (non-alkaline) flux. The resulting product, steatite porcelain, is eminently suitable for electrical insulators. Talc is also introduced in fairly large amounts in some pottery for cooking-ware, where, together with clay and possibly alumina, it forms *cordierite* ($2MgO.2Al_2O_3.5SiO_2$); this mineral has a very low thermal expansion and therefore a high resistance to heat shock.

Wollastonite ($CaO.SiO_2$), a fibrous calcium silicate, became known as a pottery raw material in 1952 when large and very pure deposits were opened up in the state of New York, United States. It fulfils the same function as steatite and had it been known before it would have taken the place of steatite because it is available in greater consistency and has a stronger fluxing action. Apart from America, deposits have been discovered in Finland and successful attempts have been made to synthesize it.

Like talc, wollastonite is used as a flux to reduce the firing temperature and it is particularly useful for eliminating moisture expansion, a problem in the manufacture of earthenware leading to "crazing".

---

* Anorthite, a calcium alumino-silicate, has already been dealt with under 2.2, "Feldspathic Minerals".

2.3.2. Carbonates

Mineral carbonates offer the simplest means of introducing alkaline earth, especially lime and magnesia, because they are available abundantly and do not contain any additional oxides, e.g. silica. (Compared with the silicates they suffer from the disadvantage of evolving carbon dioxide which may cause difficulties during firing.)

Calcium carbonate ($CaO.CO_2$) occurs as the mineral *calcite*, in limestone, marble, and chalk. It is also a major constituent of marls (mixtures of clay and chalk). In Britain and the United States, calcium carbonate is usually introduced as *whiting*, meaning finely ground chalk in Britain and ground limestone in the United States.

Magnesium carbonate ($MgO.CO_2$) occurs as *magnesite*, one of the most important deposits being in Austria. It has been used less frequently in pottery than the various forms of calcium carbonate because it is comparatively rare.

Magnesia is seldom introduced into pottery without lime and instead of adding magnesite and whiting, *dolomite* ($MgO.CaO.2CO_2$) is used. Dolomite is found in England in Derbyshire. The famous mountain range of that name in the south-eastern Alps consists of dolomitic limestone.

Barium carbonate ($BaO.CO_2$), although occurring as the mineral *witherite*, is usually obtained from *barytes* ($BaSO_4$). It is rarely used as a flux in pottery because, although it may increase translucency and fired strength, it causes excessive shrinkage and blistering. Used in larger amounts together with clay it produces "Celsian porcelain" (Herrmann, 1964), so called because its fired constitution approaches the composition of the somewhat rare mineral *celsian*, a barium feldspar ($BaO.Al_2O_3.2SiO_2$).

2.4. Miscellaneous Minerals

2.4.1. Refractory Fillers

A number of refractory materials have been used to replace crystalline silica because the latter's disadvantages (p. 27). They generally improve mechanical strength and resistance to heat shock of the fired pottery but tend to make it considerably more expensive, not only because the refractory materials are dearer than quartz but also because they require higher firing temperatures.

The most common silicate refractories used are *sillimanite* (*kyanite*) ($Al_2O_3.SiO_2$) and *zircon* ($ZrO_2.SiO_2$).

Perhaps the most important of the refractory materials introduced in pottery is *alumina* (Al_2O_3). It can be used in the form of the hydrated rock, viz. *bauxite*, consisting of the minerals *gibbsite* ($Al_2O_3.3H_2O$) and *boehmite* ($Al_2O_3.H_2O$), finely crushed and ground, or in the dehydrated form as *corundum* (Al_2O_3), an intermediate product in the manufacture of aluminium by the Bayer process. Bauxite, the largest deposits of which are in Jamaica and British Guiana, is contaminated with iron, and where white colour is important the refined dehydrated product is used.

2.4.2. *Fluxes*

Apart from the fluxes mentioned in the sections "Feldspathic Minerals" and "Alkaline Earth Minerals", compounds (usually again silicates) of lead and of boron are very occasionally used in pottery. As most of them are far more important as glaze constituents they will be discussed in Chapter 7, "Glazing" (Table 7.2, p. 137).

Suffice to mention *boron phosphate*, which gained popularity some years ago because of its exceptionally powerful fluxing action, and *barium sulphate*, which, together with clay and flint, forms *Jasper ware* developed by Josiah Wedgwood in the eighteenth century.

2.5. *Bone Ash*

Bone ash is the only ceramic raw material which is not derived from the earth; it is only used in bone china. The action of heat on bone ash, consisting essentially of hydroxy-apatite, can be expressed as follows:

$Ca(OH)_2.3Ca_3(PO_4)_2 \rightarrow 3Ca_3(PO_4)_2 - CaO - H_2O$
Hydroxy-apatite   Beta-tricalcium phosphate

Bone ash is produced from cattle bones which are reputed to be the purest. Bones from other animals, including man, are contaminated with iron oxide.

The production of bone ash involves the degreasing of bones from abattoirs by either boiling with a suitable solvent and digesting with steam or by applying special water treatment. The degreased bones are crushed and calcined to about 900–1000°C. The pieces of bone ash are comminuted by water grinding, usually in ball mills. Milling breaks up the aggregates of ultimate crystals. These are extremely fine, extending into the colloidal range, viz. less than 1 micron in diameter, thus contributing some plasticity.

The ground bone ash suspension used to be run into large containers where it was "washed" with water amounting to several times of the original slip volume to remove harmful soluble salts. After settling out the supernatant clear liquid was removed. The thickened bone ash suspension was kept for about six weeks or even longer to allow it to "age".

Unlike other ceramic raw materials, bone ash is not inert if suspended in water. During the "ageing" stage a number of subtle chemical changes occurred which considerably influenced the plastic behaviour of the bone china mixture. The "aged" bone was dried to about 10 per cent moisture; the drying used to be done very slowly on special drying kilns in such a way that air bubbles, harmful to plasticity, were eliminated. Kiln drying was no longer necessary after the introduction of the de-airing pug mill (see Chapter 3, section 2.6, p. 47) which eliminates air bubbles from the bone china mixture by virtue of the vacuum created in this machine. De-watering of the bone suspension is now done by filter pressing which, incidentally, also removes some of the soluble salts so that the long period of "ageing" is no longer necessary.

3. Temporary Raw Materials

A number of auxiliary raw materials are used in pottery which do not form part of the finished product but are, nevertheless, vital in its manufacture. The most important of these is WATER.

Water is essential in the refining process of china clays, in the flotation of quartz and feldspar, in the general preparation and mixing of the constituent minerals and in the traditional methods of shaping the actual articles.

Apart from water, the temporary raw materials used in pottery manufacture can be described under the collective term of ORGANIC BINDERS. Whereas with water the amounts involved are very high (i.e. well over 50 per cent, in the refining and preparation of the general inorganic raw materials, 33 per cent for shaping in the liquid state and 20–25 per cent for shaping according to traditional plastic methods), the quantities of organic binders do not normally exceed 0.5 per cent. They include natural as well as synthetic products, such as gum arabic, gum tragacanth, glue, cornflour, starch, wax emulsions, alginates, acrylates, celluloses, carboxy methyl celluloses, polyvinyl alcohol, etc. Synthetic binders are usually preferred, as they are more closely controlled.

The main purpose of organic binders is to influence the rheological properties of clays and their mixtures with other minerals, i.e. to improve plasticity and to increase mechanical strength of the articles prior to firing. Many of the organic binders act as flocculants (coagulants) or more often as deflocculants (important for liquid shaping—see Chapter 4, section 3, p. 71—although this process relies mainly on inorganic salts, rather than organic binders). Some organic binders have certain side effects, such as bacterial growth or frothing (mostly encountered with glazes—see Chapter 7, section 9, p. 140). In such cases the appropriate antidotes have to be introduced.

NB: Raw materials for glazes and colours are dealt with in Chapters 7 (Glazing, Table 7.2, p. 138) and 8 (Decoration, Table 8.1, p. 154) respectively.

Further Reading

General: **Worrall** (**1982** and 1986), Ryan (1978), Konta (1979, 1980), Singer & Singer (1963, pp. 3–264).
Crystal Chemistry: Budworth (1970).
Clays: **Worrall** (**1986**), Keeling (1961 and 1962).
Kaolin: Trawinski (1979 and 1981).
Ball Clay: Mitchell (1974).
Forms of Silica: **Dodd** (**1953**), Sosman (1955).
Quartz: Weiss (1979).
Feldspar: Williamson (1980).
Bone: Beech (1959), Dinsdale (1967), Forrester (1986), Webster and others (1987).

3

Preparation of "Body"

The potter refers to his material in the unfired ("green") state as "clay", no matter whether he works in a studio pottery with clay dug from his back garden or in a bone china factory with a prepared mixture containing only 25 per cent clay. When he speaks of the "clay-state" he means the ware in the unfired condition.

The technologist and the more sophisticated potter call the prepared mixture of clay with other materials the "body". The term applies to the fired as well as the green state and thus covers both the mixture to make the product and the main part of that product as opposed to the glaze.

1. Body Composition

The composition of a body depends, of course, on the raw materials available and the type of pottery one wishes to make. If it is the intention to utilize local raw materials these are first examined with a view to their ceramic suitability and based on the findings obtained the most appropriate type of pottery is chosen. Centres of pottery industry have usually been started around clay sites but not necessarily so (there are no pottery works in Cornwall which use china clay or china stone). Others have been founded near coalfields and sites of refractory clays (fireclays) suitable for the manufacture of containers in which the ware is fired. Clay and coal are often found in adjacent or even the same sites.

The composition of pottery bodies is still largely a matter of trial and error. Raw materials used to vary a great deal and the craftsman potter was able to adjust his methods accordingly. (If, for instance, the clay became less plastic he used a little more water and took extra care in preventing cracking or distortion of the pieces. Or, if the flux content of one of his materials decreased he increased the firing time or temperature.) However, the machine cannot, of course, make such adjustments and it is therefore essential for the raw materials to be consistent. Moreover, the progressive potter needs to know the chemical composition of his raw materials and of his pottery body. For purposes of comparison the ultimate chemical composition, normally expressed as the individual oxides, is not as useful as the mineralogical constitution.

The commonest types of pottery consist basically of kaolinite, quartz, and feldspathic or micaceous matter and it is these constituents in which the potter is mainly interested. They make up the so-called "rational" analysis, also referred to as the "proximate" analysis. The rational analysis cannot be regarded as an exact analysis because the figures obtained are arrived at by elimination and subtraction. To obtain an idea of the mineralogical composition it is more expedient to calculate the rational composition from the chemical analysis (see Appendix, p. 229).

2. Traditional Methods of Body Preparation for Shaping in the Plastic State

The studio potter usually buys the body ready prepared from a "miller" or supplier of pottery materials. It is supplied in polythene bags in a condition suitable for shaping his pots without any further preparation. The industrial potter may do the same but as a rule he purchases the mixed body in powder form, or more often the individual raw materials.

2.1. *Treatment of Plastic Materials (Clays)*

China clays which reach the pottery works in the purified and dried state are mixed with water in *blungers*, which consist essentially of a cylindrical or hexagonal vessel provided with a stirrer (see Fig. 3.1). The concentration varies with the degree of plasticity of the china clay and is around 30–40 per cent, which gives a free flowing if somewhat sluggish consistency. In that state the clay suspension is known as "slip" and is ready to be incorporated into the body.

Ball clays and other sedimentary clays are supplied as mined, shredded, or, in some cases, dried and pulverized. The moist clays are sometimes left to "weather" before being "blunged". Owing to the greater fineness of their particles giving greater cohesion they are less readily broken up and take longer to mix with the water than china clays.

2.2. *Treatment of Non-plastic (Hard) Materials*

The hard materials feldspar, quartz, etc., are often supplied in the ground condition ready to be incorporated into the body, but sometimes in a coarse-mesh grade which requires further comminution at the pottery works. Some factories whose products are of exceptionally high standards used to buy these materials in the as-mined condition, e.g. as hand-selected lumps to ensure highest purity and consistency.

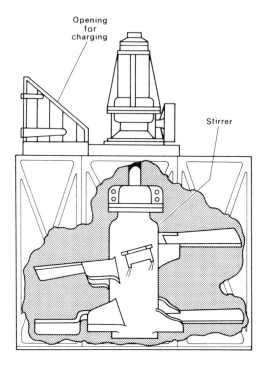

FIG. 3.1.   Blunger. (By courtesy of William Boulton Ltd.)

2.2.1. Calcination

The changes which flint and quartz undergo on heating make them more friable and they (particularly flint) are, therefore, calcined to facilitate subsequent crushing.

2.2.2. Crushing

The hard lumps are broken up in jaw crushers consisting of two serrated plates, one stationary the other moving. This produces small pieces of say up to 50 mm (2 in) in diameter which are fed to fine crushers, usually of the roller type, which reduce the size further, e.g. 10 mm (½ in). The crushers are made of manganese steel or chromium steel and their wear causes fine particles of iron to contaminate the materials giving dark specks in the finished ware. Although non-magnetic steels are harder and therefore wear less and cause less contamination, magnetic steels are usually preferred since contamination can be removed by magnets.

One way of overcoming iron contamination is to use *edge-runner mills* consisting of granite wheels ("runners") rotating on a solid or perforated granite "pan". The process is very dusty. Although contamination of the material by granite cannot be avoided this is far less harmful than iron, granite containing the same constituents as the body.

2.2.3. Grinding

The crushing operation is followed by grinding, the most important stage in the comminution of the hard materials since grain-size distribution in pottery bodies, to some extent, has a greater influence on their behaviour and properties than has composition. Before World War 2 so-called *grinding pans* represented the main means of grinding in Britain. They worked on a principle similar to the edge-runner mill, but water was used. A combined crushing and grinding action was produced by heavy chert stones running over a solid stone bottom, consisting of granite and smaller chert stones. Pan grinding fulfilled two processes: the fine crushing and the grinding. The pan could deal with material of up to 30 mm in diameter. Pan grinding provided a greater proportion of coarse grains and of very fine grains than did ball milling (see below). The wide range of particles was responsible for a better packing density which, in turn, led to improved plasticity and green strength, and lower shrinkage of the body. However, control of grain size distribution was difficult and for this reason the much more controllable *ball mills*, in conjunction with fine crushers, have replaced pans.

Following normal pottery practice, grinding in ball mills is almost invariably done with water. Suspensions of fine particles, i.e. "slips", are obtained as with blunged china clay. Ball mills consist of large steel drums lined with quartzite or ceramic blocks. Flint pebbles of various size gradings have been used as grinding media for many years and found effective, their main asset being low price.

More recently high-density balls or cylindrical rods have been introduced. They are usually of the alumina type consisting either of pure aluminium oxide or more often a ceramic body with a very high alumina content (>85 per cent).

Alumina grinding media, by virtue of their higher density, produce a greater impact. This allows grinding times to be reduced to a fraction of those required for flint pebbles and output is increased enormously. Alumina media also have a much higher abrasion resistance and hardness. For this reason they give little contamination and last much longer than flint pebbles. Alumina grinding media, although more expensive, are, therefore, more economical than flint pebbles in the long run.

To obtain highest efficiency and optimum results the volume ratios of cylinder–grinding media–grinding charge, the weight ratios of grinding charge-water and the grinding speed must be carefully selected. The volume of media including air spaces between them is about 50 per cent of that of the ball mill. Media plus material to be ground plus water occupy about 55–60 per cent or less of the volume of the ball mill. The weight of media is about 3–4 times that of material to be ground and the water content of the slip is very roughly 30–40 per cent, depending on the type of material. The grinding speed depends on the size of the mill, the larger the mill the slower the speed.

The size of media is also important; broadly speaking, the finer the material charged the smaller should be the grinding media.

Ball milling is a batch process, each charge being emptied after the grinding time required. Milling is sometimes continuous, as in the so-called *Hardinge* mills where that portion of the material which is fine enough is automatically discharged, fresh material being automatically added into the mill at the same time.

Vivian (1980), having found the motion of the media load in ball mills unsatisfactory, has developed a "*slant mill*". This consists of a barrel which is inclined at 45° to the horizontal. All the media are always active and roll continuously on the mill shell in contact with feed particles; the rate of fine grinding is, therefore, high. The output for products of equal fineness is two to four times greater in slant mills than in conventional ball mills. Moreover, power consumption in slant mills is less because the media are not lifted as a closely packed static mass as in conventional ball mills.

Another type of mill developed only fairly recently and finding use in the pottery industry is the so-called vibro-energy mill. It is particularly suitable for very hard materials and where great fineness is required. In normal ball mills the impact provided by the grinding media to break or chip particles is far greater than needed; on the other hand the number of impacts within a certain period of time is low. In the vibro-energy mill the force of impact is reduced to that required whereas the frequency of impacts is greatly increased by vibration thus ensuring much greater efficiency.

Finally, the *fluid energy mill*, also known as *jet mill* and *micronizer*, should be mentioned. It is based on compressed air supplying the grinding energy. The material to be ground is charged to two diametrically opposed air jets, comminution being brought about by impact at very high velocities (Haese, 1985). Practically no abrasive wear of the mill lining occurs with this method, which produces very fine particles in a very short time. However, its range of particle sizes is greatly limited. A wide range of grain size distribution is generally sought in pottery bodies for reasons already stated (viz. good packing density, leading to improved plasticity and green strength, and to low shrinkage).

Grain size distributions of the ground hard materials vary from factory to factory. A typical particle distribution is as follows:

+ 20 microns	30 per cent
− 20 + 10 microns	50 per cent
− 10 microns	20 per cent

2.2.4. *Sieving and Magneting*

After grinding the suspension (slip) is passed over sieves usually of the vibratory type, 80–200 B.S.S. mesh, to ensure that no extraneous substances such as wood fibre or over-sized material pass into the body. Likewise, in order to prevent small bits of iron worn off the crusher from contaminating the body, causing dark specks in the fired pottery, the slip is passed over powerful magnets (electromagnets or permanent magnets). It is then ready for incorporating into the body.

2.3. *Weighing and Mixing of Prepared Body Ingredients*

2.3.1. *The Continental European Method* (see Fig. 3.2)

The "dry" method is used on the continent of Europe and merely involves weighing out the materials in the dry state prior to charging blunger or ball mill. Due allowance has to be made for the moisture content of clays which may vary from 3 to 20 per cent.

After grinding and blunging the slips (aqueous suspensions) of materials are run into a large container called an "ark", equipped with a suitable stirring device to ensure that a homogeneous mixture is obtained.

It is also possible to dispense with the separate blunging of the clays. In this case they are charged straight into the ark and mixed with the slip of the ground hard materials.

In some cases the mixed body slip is transferred from the "mixing" ark to the "press" ark which is large enough to hold several mixings. Usually, blunged "scraps" (consisting of body returned from the potting shops, either as surplus or actual articles which were found faulty prior to firing) are also added to the mixing ark. With traditional shaping methods the amount of scraps to be recycled is quite high, i.e. one-third (or more) of the total body. Occasionally, problems arise concerning the removal of foreign bodies; therefore, extra sieves and magnets are inserted in the flow line of scraps (not shown in Fig. 3.2).

The Continental European method suffers from the disadvantage of a certain lack of flexibility. As it is practically impossible to obtain a correct average moisture content of china clay, slight inaccuracies in the proportion of raw materials are unavoidable.

2.3.2. *The English Method* (see Fig. 3.3)

This method, which is largely used in Great Britain, is somewhat complicated but ensures greater accuracy as moisture content of the raw materials does not affect it. It is more flexible than the dry method. Its disadvantage is the greater amount of floor space required. The actual slips of both plastic and hard materials are weighed and as it is impossible to keep

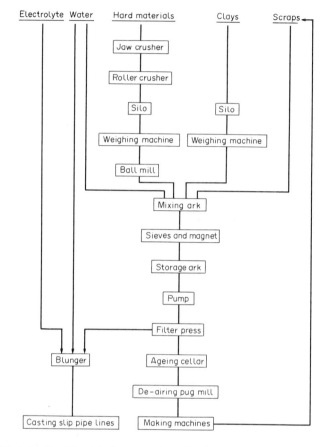

FIG. 3.2.   Traditional body preparation—Continental European method.

their water contents exactly the same from mixing to mixing, a special calculation involving volume as well as weight is applied. This is based on Brongniart's formula which allows the dry weight to be calculated from the "slip weight" in a given volume:

$$D = \frac{(W-V)g}{g-1},$$

where D = dry weight of material in volume V,

W = dry weight + water in volume V (slip weight)*,

V = volume in which W is contained,

g = specific gravity of material.

_____

* According to British convention the slip weight is usually expressed in ounces (or grammes) per pint.

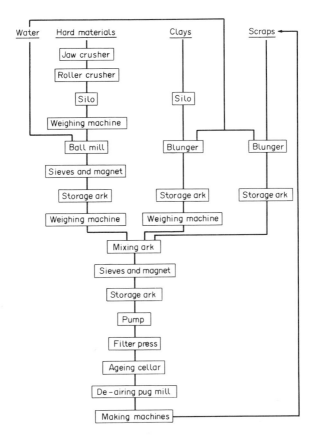

FIG. 3.3.   Traditional body preparation—English method.
Preparation of casting slip identical to arrangement shown in Fig. 3.2.

The normal solids content of china clay slip is 30–40 per cent. Weights of slip vary and in order to avoid time-consuming calculations whenever a mixing is weighed out, the required weights are read off tables which are prepared in advance, covering the whole range of likely variations in water content. Accurate weighing machines operating in large units combining weight with volume measurements are used.

In the "wet" method the ground or blunged slips of raw materials are kept in large "storage arks". From these the slips can be passed to the weighing machines any time and in any quantity—in case several bodies of different composition are required to be made. The weighed slips are run into the mixing ark to be mixed. The procedure is then essentially the same as described under 2.3.1., "Continental European method". The body slip passes from the "mixing" ark into the larger "press" ark or "finished" ark to be stored. On its way it has to travel through sieves and over magnets as did the hard materials in order to make sure that no contamination occurs. For the same reason the blunged scraps return slip is sieved and magneted (not shown in Fig. 3.3) before it passes into its storage ark.

2.4. *Filter Pressing*

The potter requires the body in the plastic state (except for slip casting (Chapter 4, p. 69) or dry pressing) when the moisture content is about 21–25 per cent. The prepared body slip, which has a water content of 50 per cent or over, has therefore to be "dewatered". This used to be done thermally on drying kilns as for bone ash. Due to the application of heat, air was expelled from the body and its plasticity was thus improved. This rather time-consuming method of dewatering gave way to mechanical removal of water by the *filter press* (see Fig. 3.4) which, in turn, is being replaced in advanced plants by more controllable machinery (see Section 3.1 below).

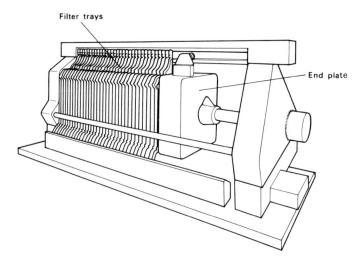

Filter trays

End plate

FIG. 3.4.   Filter press. (By courtesy of William Boulton Ltd.)

The ceramic filter presses are similar to those used in the chemical industry. They consist of a series of grooved metal trays which, when tightly pressed together, form chambers; each chamber is lined with a filter cloth. Both frames and filter cloths have a central hole so that the body slip which is pumped into the press can pass through the whole of the press filling all the chambers. Further pumping causes excess water to be squeezed out through the filter cloths. The press pump is set so that no more water is pressed out when the body has reached the correct consistency (water content of 21–25 per cent). The pumping is then stopped, the press opened up, and the "filter cakes" of body taken out or allowed to drop out of their chambers.

2.5. *Homogenizing*

The "filter cakes" as received from the filter press are not homogeneous; their moisture content is greater in the centre than towards the periphery. For this reason the filter-pressed body has to be conditioned so that it becomes uniform.

Once potters thought that this homogenizing was materially assisted by ageing the body in damp cellars.[*] There has been some controversy about the time of ageing. "Up-to-date" potters do not age their pottery body at all; others stipulate two days, more conservative ones a fortnight or even six months; up to not very long ago a small old German porcelain factory, famous for producing artistic ware, allowed three years' ageing. Marco Polo observed that in China (in the thirteenth century) the ageing time was 30 or 40 years. Father prepared the body for his children.

The body used to be homogenized by hand; the method, still used by some studio potters, is called *wedging*.

The *wedger* places a large lump of body on a smooth table, cuts it with a piece of wire in half in a vertical direction, takes one half in his hand, raises his arms as far as possible above his head and slaps the half of body lump in his hands on to the other half of body lump on the table; he does this with considerable force. He then turns the reunited lump of body through 90°, again cuts it vertically into half and forcefully slaps the one half on to the other. He goes on doing this until the texture of the body is uniform. The important point is to turn the body through 90° each time it is cut, thus ensuring intimate mixing which is usually achieved in 15 minutes.

Great force is used not only to unite the two parts and knit them closely together but also to eliminate any air bubbles from the body. Air bubbles (after iron specks the potter's biggest problem in body preparation) not only become visible as blisters in the fired ware but also impair plasticity and general workability.

Wedging well done provides a body of excellent workability, superior to conventional mechanical homogenizing as obtained with the *kneading table*, popular on the continent of Europe before World War 2 and still used in the 1960s in Japan (Allen, 1966). Hand wedging is very time-consuming and has become prohibitively expensive in industrial establishments, a long while ago because of high labour costs.

2.6. *Mechanical De-airing*

The expelling of air from the body, so vital for plasticity, is now left to the *de-airing pug mill*, usually referred to as the *"pug"* or the *de-airing extruder*. It

---

[*] It has indeed been proved experimentally that micro-organisms greatly influence the ageing process, thereby increasing the homogeneity and plasticity of clays and clay bodies (Oberlies and Pohlmann, 1958).

is an essential piece of equipment, especially for bone china which, because of its low clay content, could not be manufactured without it, now that kiln drying of bone and hand wedging are no longer employed.

The de-airing pug mill resembles a giant mincing machine with de-airing chamber and screw propulsion extruder (see Fig. 3.5). The filter cakes are shredded and mixed; the mixed body then passes through a grid into the vacuum chamber and is finally consolidated and extruded into rolls of the size (up to 200 mm diameter) required by the potter.

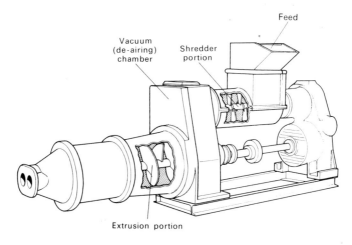

FIG. 3.5.   De-airing pug mill. (By courtesy of Edwards & Jones Ltd.)

Although universally used, the de-airing pug mill is not ideal because de-airing is not complete, and the shearing of the body leaves certain patterns usually referred to as "laminations" which lead to distortion and other faults during firing. It has been claimed that such laminations can be eliminated by special vibratory arrangements. However, extensive work at the British Ceramic Research Association (now British Ceramic Research Limited) has led to the conclusion that none of those devices removed laminations (Astbury, 1962; Hodgkinson, 1962). Towards the periphery of the body "roll" there is greater compression during extrusion than in the centre. This inhomogeneity can again cause distortion.

3. Advanced Methods of Body Preparation for Shaping in the Plastic State

3.1. Dry Preparation Involving Intensive Mixer

The traditional methods of body preparation require very large volumes of water which have to be removed again. This is wasteful. What is more important is the fact that the filter press does not allow the close control of the moisture content of the body demanded by modern shaping methods.

Different articles, made on different machines, require body of different "stiffness" and hence moisture content. Varying the pressure during filter pressing allows a certain adjustment in the water content of the body, but it is not sufficiently sensitive. Moreover, there is a lack of uniformity within the filter cakes: the peripheral portions are stiffer (containing less moisture) and the centre is softer (containing more moisture) than the average. Also, filter pressing does not lend itself to automation.

To surmount these disadvantages, *dry mixing* has been introduced. The finely ground and pulverized dry materials are mixed in a "Z" blade mixer with the exact amount of water required. The mixed plastic body then passes direct to the de-airing pug mill.

This simple method which has been in use in the United States for some considerable time has found no favour in Europe, probably because mixing in this way cannot be as effective as slip mixing when all particles are surrounded by a film of water. The particles thus have a greater attraction for each other. This gives better packing, better plasticity and green strength, less shrinkage and hence less distortion and increased reactivity at lower temperatures.

However, the development of the *counter-current intensive mixer* has made it possible to overcome the objections to dry preparation. This sophisticated piece of equipment (see Fig. 3.6) has an exceptionally high mixing intensity. Its special feature is the clockwise rotating mixing pan and the eccentrically placed counter-clockwise rotating mixing "stars". The mixing stars may be equipped with blades or kneading bars. The body particles travel over great areas (due to the motion patterns of mixing pan and mixing stars) while an independently driven, fast rotating *high-energy rotor* causes a high mixing intensity in a small area (Ries, 1973).

A flowsheet of dry mixing, incorporating the intensive mixer, is shown in Fig. 3.7 (Cubbon and Till, 1980). It will be seen from Fig. 3.7 that the automatically weighed dry materials are mixed with blunged *"scrap returns"* slip (as also shown in Figs 3.2 and 3.3) to produce the plastic body in the intensive mixer. There is no need to add water to the mixer, all the water needed for the correct moisture content of the body being supplied by the scrap returns slip. In all pottery factories the amount of scrap returns available is likely to be more than enough to provide sufficient plastic body of the precise moisture content and hence stiffness required. Using Brongniart's formula, it is easy to calculate how much returns slip has to be added.

The efficiency of the intensive mixer can be demonstrated by incorporating small amounts of a coloured organic substance, normally difficult to admix. The resulting mixed body will be completely uniform without any signs of striations, thus proving perfect homogeneity.

The dry preparation is less labour-intensive than the traditional wet

FIG. 3.6.   Counter-current intensive mixer. By courtesy of Maschinenfabrik Gustav Eirich.

preparation; labour requirements are reduced to a quarter of the latter (Nilsson, 1976). Moreover, dry preparation is faster and does not incur losses arising from spillage in filter pressing (Ries, 1979). Dry preparation is cheaper than the method involving spray drying (see below, section 3.2) as it does not require any fuel.

There is one serious disadvantage with dry preparation: the potter is no longer able to remove iron contamination by wet magneting of slips; the likelihood of unsightly specks on the finished ware is, therefore, greatly increased. Dry magneting is far less effective than wet magneting, unless high-power magnets are used, which would greatly increase the cost.

3.2. Preparation Based on Spray Drying

The principle of *spray drying* has been defined as follows (Cubbon and Till, 1980):

> A suspension of particles (slip) is projected as a cloud of fine droplets into a stream of hot gas, contained within a cylindrical chamber with a conical base. Drying by evaporation from the large exposed surface is rapid, converting the droplets into solid granules which fall from the gas stream and leave the chamber through a valve, situated in the base of the dryer. The vapour driven off is extracted from the chamber via cyclones and wet scrubbers to atmosphere.

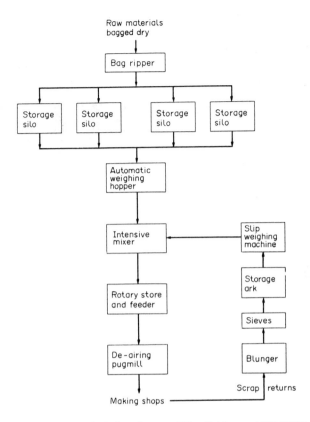

FIG. 3.7.   Dry mixing method, flow diagram. (After Cubbon and Till (1980), p. 14.)
Reproduced by permission of Verlag Schmid GmbH.

A flowsheet, illustrating the principle of the spray drying process, is given in Fig. 3.8 (Cubbon and Till, 1980). There are two main types of spray dryers, one based on nozzle atomization, the other on disc atomization; the former is used for smaller units, viz. in pottery factories, and is shown schematically in Fig. 3.9 (Cubbon and Till, 1980).

Spray drying retains the advantages of slip mixing, appertaining to the traditional methods, but eliminates the cumbersome and uncontrollable filter pressing stage. The preparation of plastic body by spray drying is invariably linked to the intensive mixer, described above. A flowsheet comprising a centralized body preparation system based on the *spray dryer–intensive mixer* method is shown in Fig. 3.10 (Cubbon and Till, 1980).

For reasons of economy, the solids concentration, i.e. the density of the slip, should be as high as possible: the less water has to be thermally removed, the less fuel is required. A certain degree of fluidity is necessary for effective spray drying. It is possible to obtain the required fluidity in a high-density slip (equal to a low-density one) by adding small quantities (not exceeding 0.3 per cent) of a deflocculant; this expedient is applied to the shaping in the liquid state, i.e. slip casting (see Chapter 4, section 3.1, p. 69).

The deflocculants normally used for slip casting tend to impart properties to the body undesirable for plastic shaping. Special liquefying agents have been developed which disintegrate in the spray drying process, so that they no longer harm the plastic body (Helsing, 1970).

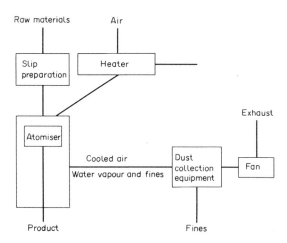

FIG. 3.8.   The basis of the spray drying process, flow diagram. (After Cubbon and Till (1980), p. 4.) Reproduced by permission of Verlag Schmid GmbH.

Compared with the traditional methods, tied to filter pressing, the spray dryer–intensive mixer procedure has all the advantages of the "dry" method, viz. very close control of moisture content (to less than 0.5 per cent) and very high degree of body uniformity and homogeneity. The latter two properties are not only a result of the treatment in the intensive mixer, but are also due to the action of thermal dewatering which does not cause segregation as with mechanical dewatering (filter pressing). Thermal dewatering also retains the finest clay particles, liable to be lost during filter pressing, and thus improves plasticity. Moreover, the spray drying route has been claimed to be more economical than the traditional plastic route, as no subsequent drying of the articles is involved; the dryers for eliminating residual moisture in the pieces after plastic shaping are generally regarded as inefficient, whereas the spray dryer provides efficient drying.

The only disadvantage of the spray drying procedure compared with the "dry" method is cost, arising from the need to use fuel. The crucial advantage of spray drying over the "dry" method is the easy removal of iron contamination by magnets through which the liquidized raw materials have to pass, thus greatly reducing the incidence of dark specks in the finished ware, dry magneting of the pulverized raw materials being far less effective.

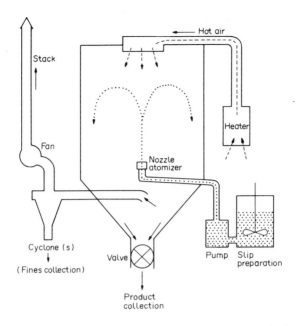

FIG. 3.9. Nozzle atomization spray dryer with counter-current air flow. (After Cubbon and Till (1980), p. 4.) Reproduced by permission of Verlag Schmid GmbH.

3.3. *The Computerized Body Preparation*

Around 1969, a large company manufacturing porcelain in the Federal Republic of Germany erected a central body preparation plant, designed not only to serve its own factories, but also to supply other pottery firms throughout the world with up to eighteen different porcelain body compositions in controlled granulate form (Dorschner and Strobel, 1970).

The body preparation revolves around the spray dryer. The plant consists essentially of a 50-m high tower which is designed to allow gravity feed throughout. Bearing in mind that numerous blungers are required, the basic construction element chosen was a hexagonal cell, corresponding to a honeycomb structure. This kind of construction makes extremely good use of all available space, and besides provides optimum structural strength. The system ensures complete freedom from dust. Only one person is in charge to look after the whole plant.

Every operation is initiated and monitored by computer, from the storage of raw materials via preparation of the slip and the spray drying process, to the final despatch of the granulated body to the user (Hohlwein, 1972). The main functions of the computer are summarized as follows:

1. Placing the raw material in the correct silo.
2. Indicating to the person in charge when the silo is full.

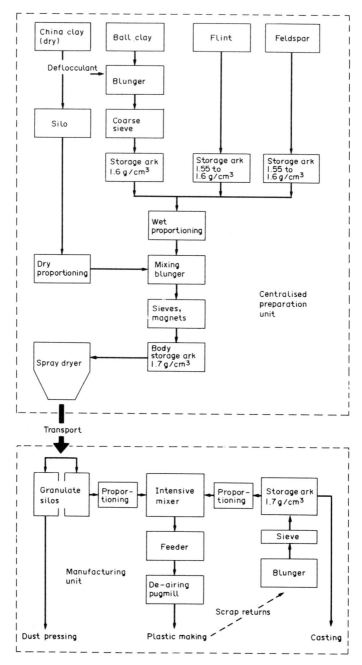

FIG. 3.10. Centralized body preparation system based on the spray dryer—
intensive mixer method, flow diagram. (After Cubbon and Till (1980), p. 14.)
Reproduced by permission of Verlag Schmid GmbH.

3. Adding the correct amount of each material to the mixing blunger for whatever body composition is required.

4. Monitoring the moisture content of each raw material, making allowance in the amount to be weighed.

5. Allowing for any change in chemically combined water in case of a change in quality of a kaolin.

6. Adding the correct amount of water, allowing for varying moisture contents of the raw materials.

7. Monitoring pH of mixed body slip, adding the required amount of electrolyte to ensure constant pH.

8. Watching over all aspects of spray drying, such as rate of feed, temperature of slip, air intake and exhaust, etc.

9. Logging major operational events and production dates, so that bottlenecks are immediately detected and remedied.

10. Determining the selling weight of prepared body granules and initiating the printouts of the despatch documents required.

At the inception of the plant a chemical analysis unit was installed, based on X-ray fluorescence, to monitor changes in the chemical composition of the raw materials.

The plant is highly flexible. It was realized that certain concepts might become out of date and new ones necessary in an effort to keep up with future requirements. Provision has been made for such contingencies, so that there will be no need to rebuild or alter the plant.

In 1986 a central body preparation plant was opened by the largest china clay producers in England (Wood, 1985). Like the German plant, it is flexible and computer controlled. Body is not only supplied in the form of spray dried granulate, but also in the form of plastic body "reconstructed by back tempering" with an intensive mixer, and in the form of suspensions of 40 to 60 per cent solids. Product consistency of a high standard, not obtainable in the past, is achieved by the most up-to-date metering systems, including "OSCAR" (On Stream Chemical Analyser and Recorder), a process developed by the company, providing ultra-rapid XRF (X-ray fluorescence) analysis.

4. Methods of Body Preparation for Shaping in the Liquid State

Complicated shapes which cannot be produced with plastic body have to be made by "*slip-casting*", involving body in the liquid state, usually referred to as *casting slip*. For various reasons (see Chapter 4, section 3, p. 69) its solid content should be as high as possible, i.e. 70–75 per cent, compatible with adequate pourability. Mixed slips prior to filter pressing have a water content of about 50 per cent; slips can be made into suspensions of almost equal fluidity, but containing only 30 per cent water, by adding small amounts of deflocculants (not exceeding about 0.3 per cent) before incorporating the extra solids necessary.

Casting slips can be prepared by passing the weighed dry raw materials to a blunger to which the correct amount of water and deflocculant has been previously added. The quartz component is usually introduced in slip form in order to avoid dust which could cause pneumoconiosis. The mixed slip may be passed to an extra mixing ark to which the final adjustments of deflocculant are added, or it may be run or pumped into the storage ark and thence to pipelines in the casting departments (see Fig. 3.2). In many factories there are three storage arks, one which is in use, a second in which the slip is being "aged" for at least 24 hours to allow it to become stable, and a third which is being filled. The casting slip in the storage arks is gently stirred (agitated) continuously with paddles or "gates" in order to prevent segregation and sedimentation. Scrap returns are blunged in a separate blunger and the resulting slip is added to the mixing ark.

Another more widely used method involves blunging broken up filter press cakes with the required amounts of water and deflocculant, followed by the general procedure described in the previous paragraph.

The blunging of filter cakes is generally preferred for reasons already indicated previously, viz. the more efficient removal of iron contamination with wet magneting which reduces the likelihood of specking in the finished ware. Moreover, the method is convenient and time saving, omitting extra weighing operations, unless different body compositions are used for plastic and liquid shaping. Also, the control of the solids content of the slip is much easier with the filter cake route. In the case of direct preparation, very careful control of water addition is necessary to ensure that the slip is not too dilute.

5. Body Preparation for Shaping in the Dry State (Dust Pressing)

Dry or semi-dry pressing is the most economic way of making plates and other flatware on a mass production scale. The media most suitable for pressing are granules (hollow spheres) which provide the ideal flow for the pressing operation, spray drying being generally regarded as the most satisfactory method for producing hollow spheres; all other methods yield an angular granulate which has poor flow characteristics and is not suitable for rapid die filling with fast acting presses.

The principle and process parameters of spray drying have been described in section 3.2 of this chapter. The flowsheet applicable for dust pressing is contained in Fig. 3.10; it will be seen that spray dried granulate is ready for dust pressing without further treatment.

The air in the hollow centres of the granulate has to be expelled during pressing. It is, therefore, desirable to minimize the "hollowness" to avoid de-airing problems in pressing.

(An Italian firm claims to have developed a "dry" method of granulation which saves 84 per cent of energy, compared with conventional spray drying. The essential feature of the process is an agglomerator, rotating between 2200 and 2800 rpm, the particles being eventually rolled into granules; the entire system is under remote electronic control.)

Further Reading

General: **Cubbon and Till (1980)**; Singer and Singer (1963, pp. 661–708).
Grinding: **Camm (1981)**; Dams (1984); Ferrari (1985); Vivian (1980).
Intensive mixer: Ries (1973, 1979).
Spray dryer: Cubbon and Till (1980).
Computerized preparation: Dorschner and Strobel (1970); Hohlwein (1972); Wood (1985).

4

Methods of Fabrication

Manufacture of pottery is intricate and entails many steps. Readers interested in making pottery themselves are advised to consult Kenny (1954).

1. Traditional Methods of Shaping in the Plastic State

In practice the term "plasticity" has a wider meaning than that given in Chapter 2 (p. 6) and is almost synonymous with "workability". In this sense plasticity embraces a number of properties and is difficult to measure quantitatively. Apart from the factors already mentioned in Chapter 2 (shape as well as size of particles, packing, state of flocculation or deflocculation, exchangeable cations), workability is influenced by ageing, surface tension, and rate of strain.

A feature common to methods of shaping where water is involved (viz. plastic forming and slip casting) is the need for the water having to be expelled again from the article made; this causes the article to shrink.

1.1. *Freehand Shaping*

Freehand shaping is now only used by peasant potters, studio potters and industrial undertakings specializing in ornamental ware.

Coil pottery has already been mentioned in Chapter 1 (p. 2). Long coils of clay or prepared body are used to build up hollow vessels such as jugs, vases, and pots of all kinds. The very large, high-tension porcelain insulators used to be made by coiling.

1.2. *Throwing*

Throwing, freehand shaping on the potter's wheel, is the most natural mechanical form of making pottery. The potter, the clay, and the wheel form a unity in which man, his material, and his machine are in perfect harmony. Objects thrown on the wheel are perhaps the finest examples of the art of pottery. To understand clay, to fathom its potentialities, the potter must first know how to throw before its secrets are revealed to him. Therefore let us dwell on this fundamental art more closely than its present application and industrial importance would warrant. A more detailed description of throwing will also help us to understand some of the other, industrially more common, methods of shaping, since the horizontally revolving wheel as used by the thrower is also the basis for the more modern mechanized methods of shaping. With these the driving force is invariably supplied by a motor, whereas the freehand potter prefers the foot-operated treadle as it affords better integration of movement and allows more convenient control over the speed of the wheel.

A lump of homogenized clay, or body, is thrown on to the centre of the revolving wheelhead (throwing the clay or body on to the wheel while this is in motion makes it hold better). The most important task is to ensure that the lump of clay or body is in the exact centre; this is done by pushing the heel of one's left hand against the side of the lump and by pulling the lump towards oneself with one's right hand, the two hands performing opposite movements. Once the pupil has mastered the art of "centring" he is well on his way to making his first simple pot. The next step is consolidation, really a continuation of the homogenizing process (by whatever method this was brought about): the thrower alternately squeezes and flattens the spinning lump, so that it is drawn up and pushed down, thus thoroughly kneading it. He then gradually hollows out the piece of clay or body between his fingers, thumbs, and palms, drawing up the sides to form the desired shape. With all operations it is important that the thrower wets his hands and the clay sufficiently to allow smooth working. Excess water makes the clay too soft.

Sometimes the clay is not thrown on to the wheelhead itself but on to a plaster bat fastened to the wheelhead by wads of clay. The pot can be taken off the plaster bat the day after turning when it has become firm. If the clay lump is thrown on to the wheelhead itself the thrown pot has to be removed from it by a wire (an important tool for cutting clay) immediately after making when the pot is still soft. It is then easily distorted but, strange as it may seem, pots which have been pulled out of shape will be perfectly round again after standing during the night. This phenomenon is due to what is known as "the memory of clay". We shall meet the memory of clay again (Chapter 5, p. 88); suffice to mention that the pot reverted to its original shape since throwing, the ideal method of potting, imparted the right type of texture to the piece.

Throwing is really a highly mechanized art. The whole body plays a part in it, each finger is an important tool. In few other crafts does the human body perform more ingeniously co-ordinated movements. The human body becomes a complex but perfect machine, a machine with a brain and one which, given sufficient practice and experience, will produce, without any guidelines, dimensional accuracy equal to that of a standard mould. The ancient Israelites knew how to do this and so does the craftsman potter of today.

1.3. *Shaping with Plaster-of-Paris Moulds*

If we were to accept the view that basket making preceded pottery and not the other way round, as some scholars maintain, the first pottery ever made was moulded and not shaped freehand. Be that as it may, practically all industrial pottery is made by moulding. The material traditionally used for moulds is plaster-of-Paris.

1.3.1. *Preparation of Plaster-of-Paris Moulds*

Plaster-of-Paris ($CaSO_4.\frac{1}{2}H_2O$), produced by calcining the mineral gypsum ($CaSO_4.2H_2O$) at 160°C, is supplied to the pottery in powder form. It is mixed with water in the ratio of 75–85 parts by weight to 100 parts plaster. It is important to add the plaster (by sprinkling) to the water and not vice versa. A soaking period used to be allowed with blending by hand to ensure good dispersion of the plaster in the water, whereas in the plaster blending machines this is achieved merely by vigorous mixing.

In the majority of industrial potteries the preparation of the plaster slip is automated. The plaster/water ratio is closely monitored; the stirring time is strictly controlled and so is the—very short—time the plaster slip is allowed to stand, before it has reached the correct consistency for pouring. Very soon after pouring it sets and becomes solid and hard. When the plaster has set the uncombined water evaporates and leaves voids (pores). The higher the proportion of water the greater the porosity of the mould but also the lower its strength and its useful life.

The making of the moulds follows an alternating series of positives and negatives. The original model, say that of a jug, is made by turning a solid plaster-of-Paris block. From that a negative mould is made. This is split into two or more parts and as such becomes the so-called "block mould". Further plaster casts are made from the block mould, again in two or more parts and these positive casts are known as "cases". An almost unlimited amount of actual "working moulds" can now be cast from these cases. The working moulds are, of course, negative to produce the positive pottery article.

In order to increase the lives of block moulds and cases the amount of water mixed with plaster is reduced to a minimum, thus giving lowest porosity and hence strength and durability. (Working moulds of such high plaster/water ratios would be unsuitable as the articles tend to stick to them and cannot be unmoulded when dry.) The plaster moulds have to be dried before they are used in order to allow the water, not chemically combined, to evaporate. This has to be done slowly and carefully—ideally in air at room temperature. 45°C is the maximum temperature tolerable; at higher temperature there is the danger of "burning" the plaster. However, the temperature in the mould dryer can exceed 45°C in the initial stages, since the high moisture content, then present in the mould, and the high humidity in the dryer will ensure that the temperature to which the actual moulds are exposed is less than 45°C.

In an effort to increase the lives of block moulds and cases, plaster has been replaced in block moulds by polyvinyl chloride and synthetic resins such as phenol formaldehyde and various types of epoxy resins. These have not only the advantage of much increased wear resistance but, being flexible, allow easier parting and, with complicated shapes, make it possible to reduce the number of mould parts required. Furthermore, they do not need parting agents such as oils and soft soap necessary with plaster cases to prevent sticking of the fresh plaster cast to the cases. This sticking is due to the slight swelling which occurs during the setting of the plaster. Incidentally, this very slight setting expansion has the advantage of allowing the shape of the model to be reproduced accurately. To strengthen the actual working moulds and thus increase their lives they are backed by metal.

The chief advantage of plaster-of-Paris as a mould material is its low price. However, working moulds in plaster wear out fairly quickly and for high-quality ware about 100 uses is about the maximum. Besides, they have to be dried between uses (since some of the water in the body is absorbed by the plaster) and, therefore, require large storage space.

Working moulds for certain special shaping processes, such as "ram" pressing (see section 2.2, p. 66), are exposed to excessive wear and, therefore, have to be strengthened to avoid undue mould loss. This is done by subjecting the hemihydrate to a preliminary heat treatment; moreover, special additions are made to the plaster slip which deflocculate it, retard its setting and reduce its swelling. After the plaster has set, the mould is aerated with compressed air (Gawlytta and others, 1981).

Efforts have been made to replace plaster by porous sintered metals. Copper sinters readily to give a suitable porosity and pore structure. However, copper is rather soft. Sintered glass or sintered glass in combination with a sintered refractory, such as alumina, has been suggested. (Sintering is usually associated with densification in ceramic phraseology—see Chapter 9, p. 92—however, the term "sintered glass" or "sintered metal" implies porosity.) Porous ceramic moulds have successfully replaced plaster moulds, to a limited extent, in the "ram" pressing process.

1.3.2. *"Jiggering" and "Jollying"*

Pottery technology is full of quaint expressions, and "jiggering" and "jollying" are the somewhat comical names given to the general pottery manufacturing methods used industrially up to about 1950 and still applied today, on a limited scale, for "difficult" pieces in "difficult" bodies, like large bone china dinner plates (exceeding 250 mm diameter).

A slice of extruded body, ex de-airing pug mill, is placed on a horizontally revolving plaster-of-Paris mould which forms one side of the article, while a stationary tool (template) shapes the other side. If, as with plates and saucers, the mould forms the face and the tool the back, the process is called "*jiggering*". If, as with cups, the mould forms the outside and the tool the inside, the process is called "*jollying*".

When high quality was required the jiggering operation was preceded by a separate operation, known as "*bat*" *making*, whereby the slice of body was flattened on a revolving plaster head with a simple tool, to beyond the diameter of the plate. Likewise, roughly preshaped cups, so-called "linings", were hand "thrown" and placed in the cup mould before jollying started.

2. Advanced Methods of Shaping in the Plastic State

2.1. *Forming by Roller Machines*

This type of plastic making was developed very soon after World War 2 by a British firm. Continental European companies (mainly German) manufactured roller machines, first under licence and then perfected them for hard porcelain. The essential difference between roller making and the traditional methods is the use of a heated rotating die instead of a fixed template. Both die and mould revolve in roller making. Bat and lining are generally dispensed with. The principle of a plate roller machine is shown in Fig. 4.1, that of a cup roller machine in Fig. 4.3. Automatic plants are illustrated in Figs 4.2 and 4.4 respectively.

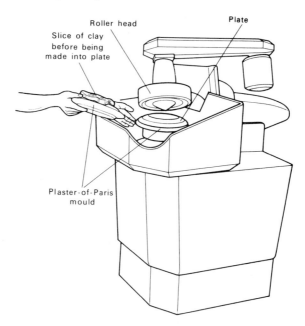

FIG. 4.1.   Plate making by roller machine. (By courtesy of Service (Engineers) Ltd.)

The success of roller making depends on a multitude of factors, i.e. the correct revolving speeds of mould, spindle and roller head, the tilting movement and dwell of the die, the degree of vacuum applied to the mould, the shape and permeability of the mould, the intensity of the water spray, body conditions, such as plasticity and water content, etc.

FIG. 4.2.   Automated plant of plate making by roller machine. By courtesy of Service (Engineers) Ltd.

The roller machines have revolutionized industrial potting, since the potter is no longer in control, as he was with jiggering and jollying, when he would use more or less water and apply more or less pressure, according to fluctuations in body consistency. The roller machine is incapable of such adjustments, hence the need for close control of moisture content of the body, as guaranteed by the advanced body preparation method, involving spray dryer and intensive mixer. (Of course, it would not be impossible to build into the machine a well-programmed microchip which would take care of fluctuations in the body, but so far it is cheaper to control the body than to equip the machine with a computer.)

The establishment of the roller machine brought certain problems. It is perhaps appropriate to mention two because their solutions offer an insight into the rheological characteristics of pottery bodies.

The first serious trouble was misshapen fired plates, made of "difficult" bodies, viz. hard porcelain. This was solved by re-introducing the making of

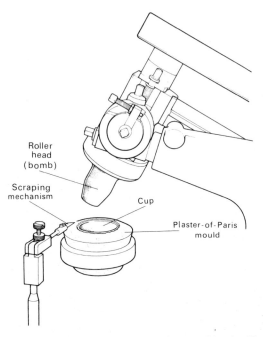

Roller
head
(bomb)

Scraping
mechanism

Cup

Plaster-of-Paris
mould

FIG. 4.3.   Cup making by roller machine. (By courtesy of Service (Engineers) Ltd.)

bats (smaller in diameter than for jiggering). The roll of body ex de-airing pug mill is not uniform. Due to the screw construction, the centre of the roll is softer, less consolidated than the portion nearer the periphery; this can be demonstrated by an "S" crack in the centre of a slice, cut off the body roll and dried very quickly. This lack of uniformity is responsible for the fact that plates made from bodies liable to develop a high amount of melt during firing, i.e. hard porcelain, tend to lose their shape. If the slice of body is subjected to the "batting" operation, the greatest pressure is exerted in the centre of the slice, thus consolidating the bat in the centre and evening out the differences in consistency over the diameter of the slice. The resulting bat is of uniform texture and thus guarantees straight plates.

The fault most common in roller making, affecting all bodies, again concerning mainly plates, is "*stretch marking*" or "stretched face". This consists of small cracks and folding marks, due to air being trapped while the heated die spreads the body slice over the mould surface. The fault is particularly pronounced with bodies of low plasticity, especially bone china, and with large size plates (most pottery faults being worse with large pieces). The movement of the body slice is governed by the following factors:

(a) the vacuum applied during shaping;
(b) the thickness of the centre section of the mould; and
(c) the air permeability of the mould.

The greater the air permeability of the mould, the less likely is the development of stretch marks. Mould permeability is not dependent on porosity. Merely making the mould highly porous by a low plaster/water ratio does not solve the problem. The most important factor influencing air permeability of moulds is the stirring time of the plaster slip prior to pouring; the shorter it is, the higher the air permeability. Short stirring times cause large crystals of calcium sulphate dihydrate to be formed; these tend to bunch and thus give a highly permeable microstructure (Cubbon, 1982).

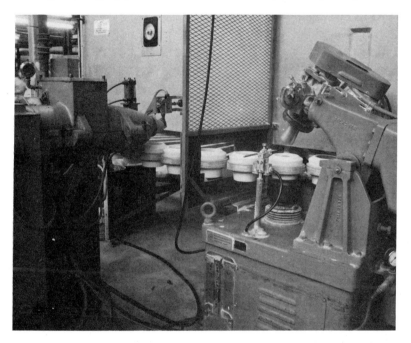

FIG. 4.4.   Automated plant of cup making by roller machine. By courtesy of Service (Engineers) Ltd.

Air flow through a plaster-of-Paris mould does not only depend on its permeability but also on the overall length of the path through the mould. Therefore, the centre section of the mould should be made as thin as possible, compatible with adequate mechanical strength (Bradshaw and Gater, 1980). A high vacuum is desirable, i.e. not less than 500 mm, especially at the start of the operation, when the roller tool contacts the slice of body, stretched face being induced at that initial stage (Bradshaw and Gater, 1980).

Oiling of moulds has proved beneficial in avoiding stretch marks. However, this extra remedy should not be necessary if high initial vacuum, low mould thickness and especialy high mould permeability are maintained.

Making cups on roller machines has hardly presented any problems; a certain degree of air permeability is desirable to prevent air being trapped between the bottom of the cup and the mould.

Roller making does not require skilled potters, but offers scope not only for the skilled fitter and engineer but also for the mechanically-minded operator who will be able to do his own adjustments to the roller machine.

The two great advantages of roller making over the traditional methods are:

1. It speeds up the process considerably; for instance, the output of bone china cups could be increased sixfold.
2. It lends itself to automation.

Automation is not confined to the actual shaping process but embraces operations like cutting the extruded rolls of body to the required length, transporting them to the roller machine, cutting off slices and placing these in the moulds, scraping off spare body (scraps being automatically returned to the body preparation plant), moving the moulds with the pieces to a pre-dryer, automatic unmoulding, final drying, removing seams and carrying out other finishing operations (see Chapter 5, section 2, p. 89), stacking (where applicable) and placing in kiln. The automated processes have been described in detail for cups (Waye, 1973) as well as plates (Niffka, 1971; Pfuhl, 1976). In some automated plants two cups of different shapes or two saucers of different shapes, or one cup and one saucer can be made simultaneously on the same machine; the saucers are made in the same manner as the cups, but the roller is stationary and unheated.

The most significant aspects of automation are the small floor space required, the very small number of plaster moulds necessary and the speed of operations, especially drying (see Chapter 5, section 1.3, p. 85).

Compared with the more sophisticated method of dry pressing, roller making has the advantage of greater flexibility. It takes little time to change from one shape to another, so that production flow is hardly disturbed.

Roller making has been confined to round shapes so far, but a roller machine for oval tableware has now been developed by an Italian firm (Anon., 1985).

2.2. *Plastic Pressing or "Ram" Pressing*

Oval shapes are still made by jollying or slip-casting. A method known as the "RAM" process (Gould, 1942) was developed in the United States for shaping oval, oblong or otherwise difficult articles by pressing in the plastic state. It is cheaper than the traditional methods because of increased output.

The plastic body is pressed between two permeable plaster-of-Paris dies. The plaster is reinforced by metal to give added strength. The dies contain woven cotton tubing which allows air to be blown through, to cause mould release of the piece and to extract water from the dies (Ott, 1975). In some cases a vacuum system is incorporated for quicker, continuous dewatering (Kieffer, 1979).

Plaster has been replaced as a die material by ceramics, such as porous sintered alumina which lasts very much longer (Skriletz, 1977), producing 75,000 pressings, compared with 600–1000 fills with plaster moulds which is sufficient for only one day's production. However, blocking of the pores of the alumina dies occurs early in the die life, causing sticking of the ware (Cubbon, 1982).

The Ram process does not always allow high-quality ware to be produced, pieces showing flow marks or becoming distorted. Distortion can be surmounted by preshaping the body slice which, however, decreases the economic advantage of Ram pressing (Cubbon, 1984).

2.3. *Impact Forming*

This process, developed by a British engineering firm for forming cup handles, is based on the ability of plastic body to withstand extensive deformation—brought about by impact—without rupture. A slug of body is forced through an orifice (by impact) into a steel die. A second steel die which is heated provides release (see Fig. 4.5). This alternative to slip casting is very fast, as the impact stroke takes place in a fraction of a second (Bradshaw and Gater, 1980).

FIG. 4.5.   Impact forming of cup handle. By courtesy of Service (Engineers) Ltd.

2.4. *Injection Moulding*

Injection moulding, well known in the plastics industry, was used for producing ceramic sparking plug insulators. The ceramic body first acted as a filler in a thermoplastic operation. The process was then reversed: the thermoplastic formed was destroyed by slow heating; the shape formed was then treated like a normal ceramic article. This process proved too expensive for making pottery articles, as the costly resin was not recoverable and as losses due to misshapen ware were unacceptably high. Thirty per cent of the volume of the shape produced consisted of resin and this led to excessive shrinkage and hence distortion.

A German firm developed a method of *ceramic injection moulding* which did not imply the use of thermoplastic resins, but relied on the natural plasticity of the pottery body (Zeidler, 1972). This process was intended as a more economic alternative to the slow production of cup handles by slip casting, as in the case of impact forming. It was attempted to fabricate seamless handles to make the process economically worthwhile.

The process comprises the injection of a slice of body into a mould passage through an orifice; the cross-section of this orifice is smaller than the smallest cross-section of the mould passage. The mould is opened to allow removal of the formed handle. Grippers rotate the handle and apply it to a waiting cup.

Another type of injection moulding (Handke, 1974) can be applied to the manufacture of plates and other flatware, as well as to shapes normally produced by Ram pressing and solid casting, but not to holloware. In this technique a body slice is introduced into a closed mould through a nozzle at very high speed. The impact of the body on the water-permeable walls of the moulds causes a phase separation between the solid particles and their carrier, the water being led away by the capillaries of the mould. The article is thus dewatered uniformly in contrast to normal drying through heat which, almost invariably, leads to stresses which, in turn, are responsible for crooked ware.

The moisture content of the dewatered article, when removed from the mould, is 12–14 per cent, the original body slug containing 22–25 per cent water. In that stage shrinkage has ceased (see Chapter 5, section 1.1, p. 84) and stresses no longer occur. The special uniform dewatering caused by the high speed of injection is responsible for producing a stress-free article. Moreover, the injection moulded piece is more compact and, therefore, shrinks less than a traditionally formed article—another reason for freedom from faults like crookedness.

Special durable, high-strength plaster-of-Paris moulds, to stand up to the injection impact, have been developed. Injection moulding requires only one single mould; this means a substantial saving in floor space, normally required for traditional shaping methods. A change-over from one shape to another, or from external to internal shaping, requires little time. There is no spillage. No first stage dryers are required.

Ten pieces per minute can be shaped by injection moulding, thus twice as many as with solid casting; injection moulding, however, is slower than roller making. It is suitable for being incorporated in a fully automated conveyor belt production line, but its use is less widespread than expected.

3. Traditional Method of Shaping in the Liquid State

This method, of ancient origin but used in Europe only since about 1730, is known as "slip casting" and is applied to figurines and shapes such as tea pots, coffee pots, jugs, and other hollow ware and sometimes to large dishes. The method is somewhat slow, involves rather a lot of space, and hence is only used where the mechanized methods of shaping in the plastic state prove difficult or impossible. Although the volume of slip-cast ware is much smaller than that produced by plastic-making methods, slip casting is treated in greater detail as its mechanism is somewhat involved. Besides, whereas the complex rheological behaviour of bodies in the plastic state is by no means fully understood, much more is known about the principles which control casting behaviour.

3.1. *Slip-casting Technique*

The process of slip casting involves pouring the slip into a plaster-of-Paris mould of the desired external shape, e.g. a tea pot or vase, until full. The water from the slip is partially removed by the pores in the plaster due to capillary action. As a result a solid layer of body—the actual cast—is built up on the inner walls of the plaster mould. As more water is absorbed the level of the slip in the mould sinks and the mould is topped up with more slip. The solid layer becomes thicker with time, although the rate of build-up becomes slower as the solid layer (cast) itself acts as a barrier. There is thus a practical limit to the thickness obtainable. After a certain time, say one hour, no more build-up occurs. The maximum thickness obtainable depends on the composition of the body and does not normally exceed 10 mm (½ in) in pottery.

When the solid layer has reached the required thickness (for instance, 3 mm (⅛ in) for a tea pot which may occur between 10 and 20 minutes depending on the composition of the body) the surplus slip is poured off and can be used again. The mould is left upside down in a slanting position for a time to allow the cast to drain. Mould and cast are then dried. During drying, whilst the cast remains in the mould, some of the water in the cast evaporates and as a result the cast shrinks. At the same time it becomes sufficiently rigid to be taken out of the mould. The remainder of the water evaporates during the final drying of the cast in air after removal from the mould.

Multiple moulds are almost invariably used so that the casts can be removed from the moulds. If the pieces (excepting very tall vases) were of a shape which could be made with one single-part mould, the cheaper method of jollying or roller making would be applied. With jugs, tea pots, etc., the sides of the moulds are usually split, giving two equal halves, whereas the bottom is supplied by a third part. As a rule there is a fourth part resting on top of the two splits acting as a "riser" so that there is no need for topping up the mould with further slip. These three or four parts of one mould form the "main body". Spouts and handles are normally cast separately, again in split moulds, and have to be stuck on afterwards (see Chapter 5, p. 90).

Figurines often require many moulds because of numerous "undercuts" which otherwise would make unmoulding difficult or more often impossible. The original model therefore has to be cut up into a number of pieces which have to be cast separately; each mould may consist of several splits. Great skill is required in working out the minimum number of moulds required. The various parts are again stuck together after unmoulding and partial drying, in the same way as handles and spouts are fixed to the "bodies" of tea pots.

Slip casting is largely mechanized today. Continuous production on a conveyor belt with automatic filling and emptying of the moulds and other modern techniques are used.

Certain parts which do not have to be hollow are "double cast" as opposed to "open cast" described so far. Double casting is applied to small parts of figurines, to handles for cups as well as to oblong dishes. With double casting the slip is applied through a casting hole in the mould and the cast is formed by two plaster surfaces—the mould outlining the shape of all surfaces—so that the whole mould space is filled with the cast. Open casting provides equal thickness throughout the piece, whereas with double casting it is to some extent possible to vary the thickness.

3.2. Formation of Slip

Casting slips are not usually produced by just mixing the finely divided raw materials with water. It is possible to use a plain water slip but the water content of such a slip has to be as high as 50 per cent. This has two very undesirable consequences:

(a) The moulds very soon become saturated and therefore ineffective. They have to be dried completely after each cast has been removed and their useful life is greatly shortened.
(b) The cast has an extremely high shrinkage which leads to distortion and other faults.

For these reasons it is desirable to have as low a water content as possible. This also explains why the potter does not use the ready-mixed slip prior to filter-pressing for casting but goes to the trouble of filter-pressing and reblunging it.

A substantial reduction in water content, down to 30 per cent or even 25 per cent and less, can be achieved by the addition of very small amounts, not usually exceeding 0.2 per cent, of *deflocculants* which make the slip fluid. Deflocculants were first used in the porcelain factory in Tournai in 1780 and remained a secret of French pottery works for over 50 years (Litzow, 1982).

By the middle of the last century sodium carbonate and sodium silicate were known to make the slip more fluid and these reagents are still widely used today. The effect of deflocculants depends to a large extent on the clays and other raw materials used and optimum type and amount have to be ascertained empirically. Often a mixture of two deflocculants gives best results.

Casting slips containing inorganic deflocculants, especially calgon (sodium hexa-metaphosphate), but also the commonly used sodium silicate and sodium carbonate, suffer from lack of stability which manifests itself in fluctuations of slip viscosity; this causes difficulties in the casting operation. Ageing of the slip before use greatly alleviates this problem. As with plastic body, there are conflicting opinions about the duration of ageing. Twenty-four hours' standing is regarded as the minimum. It has been reported that slip stability is only achieved after 5–6 days (Blasius and others, 1984).

Synthetic organic polymers, soluble in water, mainly polyacrylates, based on sodium and ammonium, were developed which caused much improved slip stability, making ageing superfluous. Other advantages were higher solids content of the slip and less likelihood of distortion of ware (Alston, 1975).

A very simple experiment demonstrates the effect of deflocculants: if clay is mixed with just sufficient water to form a smooth but very stiff paste and a few millilitres of calgon (sodium hexa-metaphosphate) or water-glass is then added, a free-flowing liquid is produced on stirring. What is the cause of this remarkable phenomenon?

3.3. The Mechanism of Slip Formation

A superficial account of the scientific principles of slip casting must suffice in the present context.

The basic factors governing plasticity, viz. the presence of colloids in clay and their cation exchange capacity, are also responsible for castability.

In a suspension of clay the balancing cations (referred to in Chapter 2, section 1.1.1, p. 10) remains close to the negatively charged colloidal particles (provided that there is no electrophoresis) and form a layer around them (Fig. 4.6). According to Gouy-Freundlich's theory, now generally accepted, there are two layers, the inner one, consisting of negative charges derived from OH^- or O^{2-} ions in the clay lattice, and the outer one being more diffuse, containing the positive ions. This concept is known as the "diffuse double-

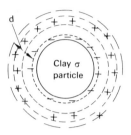

FIG. 4.6.   Distribution of charges on a colloidal particle of clay. (After Worrall
(1982), p. 30.) Reproduced by permission of The Institute of Ceramics.

layer" theory. A schematic representation of the double layer effect is given
in Fig. 4.7.

The two layers form a spherical condenser with two concentric charged
plates. The electrical potential of the system, known as the ζ *(zeta) potential*, is
given by:

$$\zeta = \frac{4\pi d\sigma}{D}$$

where σ is the amount of electrical charge per unit area on the particle,
   d is the effective distance apart of the positive and negative layers, and
   D is the dielectric constant of the surrounding liquid (water).

Increased charge or thickness of the double layer increases the zeta
potential. Alkalis break down the large aggregates and set free some of the
mechanically held water associated with them. This causes deflocculation and
hence reduction in viscosity.

If the zeta potential is increased by excess alkali, hydration of the dispersed
particles occurs. This causes an increase in the volume of the dispersed
material, and hence increased viscosity.

This is in agreement with practical experience in that excess deflocculant
often produces a thickening of the slip.

Strong alkalis, such as caustic soda (NaOH), are difficult to control since it
takes very little to upset the balance and a minute "excess" results in an
increase in viscosity. For this reason caustic soda is not used by itself to
deflocculate clay slips with the exception of some ball clays. Such clays
contain *protective colloids* in the form of lignates, humates, etc., which greatly
minimize the effects of excess alkali.

When there are no protective colloids present in the natural state,
particularly in the purer clays such as china clays, this deficiency is made
good artificially by the addition of humic acid, tannic acid, etc. In such cases
the initial alkali is introduced in the form of sodium tannate or more often
sodium silicate (containing colloidal silicic acid), the deflocculant most
extensively used with English china clays.

Protective colloids not only cause a very remarkable widening of the "safe" range of alkali additions but also bring about a decrease of viscosity. The amounts of colloidal sodium salts added are therefore lower. It is desirable to keep the addition of deflocculants as low as possible as they attack the plaster, causing a decrease in mould life. Because of their tendency to migrate to the surface of the cast pieces in the early stages of drying, they also produce faults in the ware.

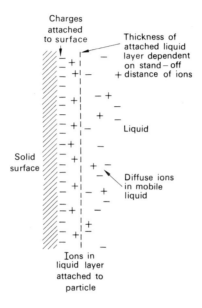

FIG. 4.7. Schematic representation of Gouy-Freundlich electrical double layer. (After Holdridge (1953), p. 71.) Reproduced by permission of The Institute of Ceramics.

The clays and water used in the preparation of slips may vary in composition and hence the optimum deflocculant addition varies. The properties used in controlling casting slips are slip density (solid/water ratio), viscosity (or its reciprocal, fluidity), and *thixotropy* (in the present context the phenomenon of the slip thickening on being left standing but reverting to its original viscosity when stirred). A control system widely used dispenses with the determination of the density of the slip and is confined to two readings on a torsion viscometer, one of the freshly stirred slip and the other after it has been allowed to stand for a predetermined length of time, usually 1 minute or 5 minutes. The difference between the two readings is a measure of the thixotropy of the slip. Each batch of slip is adjusted to those values of viscosity and thixotropy which have been established to give satisfactory casting; this can be affected by additions of deflocculant, water, or clay (body).

3.4. *The Slip/Plaster System*

In the formation of a cast the removal of water from the slip by the capillary action of the pores in the plaster has been widely regarded in the past as the only factor responsible.

However, recalling the exchange capacity of clays it seems reasonable to assume that there is an ion exchange between slip and plaster. Migration of Ca^+ ions from the plaster into the slip has been experimentally confirmed by using radioactive calcium as a tracer. Slawson (1948) showed that a mould made of earthenware of the same porosity and pore structure as a plaster mould gave a very flabby cast which completely lacked rigidity. If, however, the earthenware mould was impregnated with a small proportion of soluble calcium salt normal rigid casts were obtained.

It was proved that a cast could be produced entirely without the de-watering mechanism. A thin layer of plaster of only 0.25 mm obtained by

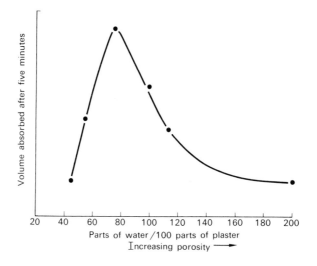

FIG. 4.8. Comparison of casting rates for different plaster/water ratios. (After Walker and Dinsdale (1959), p. 147.) Reproduced by permission of British Ceramic Research Ltd.

lining a glass beaker with a calcium sulphate film produced a cast of 2.5 mm thickness; the cast was thus ten times as thick as the "plaster mould" which could not possibly absorb the amount of water producing the cast.

Chemical reaction with casting slips containing sodium deflocculants has furthermore been shown by the occasional appearance of a white fluff of sodium sulphate on the outside of old moulds.

It would appear that there is enough experimental evidence to prove that casting is governed by a chemical as well as by a physical process.

The rate of casting depends to a large extent on the density of the plaster mould. The less water used in preparing the mould the denser it is. With normal pottery moulds made from a ratio varying between 75 and 90 parts of water to 100 parts plaster, the casting rate increases with increasing density of the mould (see Fig. 4.8). This is against expectation since water penetrates a more porous mould faster than a denser one. However, the cast as well as the mould offers a resistance to the flow of water. The relationship between the resistance of the cast and of the mould has been examined.

The results of a series of moulds of different porosities are shown in Fig. 4.9. The sum of the resistances of the mould and of the cast, representing the total resistance of the system, shows a minimum value with a mould of a plaster/water ratio of 100/75. This corresponds to the maximum rate of water absorption shown in Fig. 4.8. With more porous moulds (the types used in pottery) the resistance of the cast controls the flow; the rate of flow thus increases with decreasing mould porosity. With the less porous moulds, mould resistance is the operative factor in controlling the rate of flow, which in this case decreases with decreasing porosity.

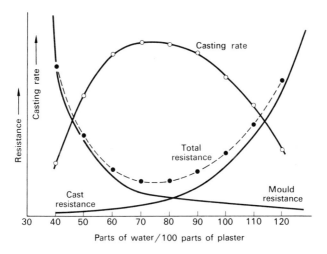

FIG. 4.9.   Combined effects of mould resistance and cast resistance upon casting rate. (After Walker and Dinsdale (1959), p. 147.) Reproduced by permission of British Ceramic Research Ltd.

Dinsdale (1959) demonstrated these phenomena by a film of an elegant experiment; he reasoned that the actions taking place in plasters of different densities could be simulated by tubes of different diameters inserted in a beaker of coloured water: the rate of flow was very much greater in the wider tube because it had less resistance to flow. If, however, a fixed resistance was attached to the tubes, the flow in the wide tube was much slower than in the narrow tube. Thus the resistance of the tubes ceased to have any influence. The fixed resistance corresponding to the cast became the operating factor.

Possibly the increased ionic migration arising from the denser moulds may also have a bearing on the faster casting rate.

Increase in casting rate in desirable but if brought about by denser moulds this advantage may be outweighed by the disadvantages: dense moulds tend to make mould release of the cast difficult, cause pinholes in the cast and have other undesirable side-effects in casting. As a rule, moulds for cast ware are therefore less dense than those for ware made in the plastic state. As a result they do not last so long as those for plastic shaping, e.g. roller making; they are also more prone to corrosion. It has been proved experimentally that corrosion of plaster moulds for casting was due to chemical, not physical, attack. Plaster is slightly soluble in water and is attacked by certain deflocculants, used in casting slips.

4. Advanced Methods of Shaping in the Liquid State

4.1. *Pressure Casting*

The advantages of casting under pressure over traditional casting are:

(a) higher productivity due to very much shorter casting times,
(b) more efficient dewatering of casts, hence:
(c) shorter drying time, and
(d) denser cast, hence:
(e) improved green strength, and
(f) less drying and firing shrinkage, hence:
(g) improved (tighter) size tolerance of ware, and
(h) less distortion of ware during firing,
(i) elimination of "pin-holes" in the glazed ware.

Despite attempts at de-airing, conventional casting slips are never entirely free from air bubbles. If these are near the surface, the glaze during the glazing firing dissolves the thin layer of body covering the bubble, thus exposing it. On cooling, the burst bubble is only partially healed. The fault thus produced is described as a "pin-hole". The great merit of pressure casting is the absence of air bubbles and hence of pin-holes.

Pressure casting can only be applied to two-part moulds and this limits it to simple shapes. Its development has not been without difficulties. Some investigators found that sticking of the cast to the plaster mould made pressure casting impractical. Replacing plaster as a mould material by an undisclosed inorganic substance/plastics aggregate removed the problem. The crucial point which emerged was the need for a sharp separation of solid particles from the water in the slip; this was achieved by moulds with extremely fine capillaries (Olgun and others, 1970).

Other research workers found the short mould life to be the main obstacle to the successful industrial application of pressure casting, ordinary plaster moulds providing only 400–700 fills (corresponding to one day's work). A British company overcame this drawback by developing porous plastic dies, based on sand/resin mixtures. However, these dies became blocked after 400 fills by penetration of the finest of clay particles (Cubbon, 1984).

The difficulties of pressure casting revolving round the mould (die) material seem to have been solved by a Swiss firm manufacturing sanitary ware, in conjunction with a German supplier of ceramic machinery. The system developed is based on special two-part dies made of an undisclosed material with outstanding filtering properties and very long service life (Luchs, 1985).

The remaining great disadvantage of pressure casting, compared with conventional slip casting, is the high capital cost of the complicated equipment, particularly the special dies. This means that, despite the considerable advantages listed above, pressure casting will be confined to items of very long production runs. Pressure casting can be applied to shapes which lend themselves to Ram pressing. The quality of pressure cast ware is greatly superior to that of Ram pressed one.

4.2. Casting by Electrophoresis

Shaping by electrophoresis makes use of the discovery that particles in a suspension migrate in a certain direction if under the influence of an electric field. When suspended in water, particles of most ceramic materials carry a negative charge. If two electrodes are dipped into an aqueous suspension of ceramic particles and a d.c. potential is applied, the particles move toward the anode and are deposited thereon. This phenomenon has been utilized for dewatering suspensions and for ceramic coatings, including deposition of ceramics on metals, i.e. vitreous enamels. It has also been considered for speeding up, and indeed replacing, conventional slip casting.

The fabrication of a simple shape (crucible) by electrodeposition is demonstrated schematically in Fig. 4.10, using sintered bronze electrodes in

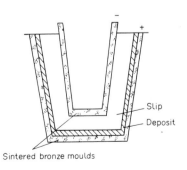

Sintered bronze moulds

FIG. 4.10. Electrodeposition of a crucible using sintered bronze moulds. (After Ryan and others (1981), p. 46.) Reproduced by permission of The Institute of Ceramics.

the shape of moulds. Other porous, electrically conducting mould materials tried included graphite/portland cement (fine-grained), graphite having been replaced by petroleum coke to give moulds of greatly improved hardness (Ryan and others, 1981).

The following advantages of electrophoretic deposition (ED) have been put forward:

(a) ED is about twenty times faster than traditional slip casting. There is a linear relationship between deposition time and deposit weight, contrary to conventional casting where the rate of deposit decreases with time (deposition ceasing altogether after a certain point).

(b) ED is self-limiting; the current stops when the correct thickness has been reached; hence much closer dimensional tolerances were forecast with ED, compared with traditional slip casting.

(c) ED, being unaffected by the viscosity of the slip, does not require close control of its rheological properties, deflocculants hardly influencing the rate of deposition.

(d) ED is cheap as very little current is used.

Serious disadvantages were encountered:

(a) Electrolysis due to migration of ions from the water to the electrode caused gases to be evolved at the electrode, gas bubbles becoming trapped in the deposit; this led to the formation of pin holes in the fired ware.

(b) Electrolysis could also have been responsible for the formation of hydrochloric acid which attacked the electrode metal, thus bringing about contamination in the form of dark specks on the fired pieces.

(c) Although the rate of deposition is very fast, less water is removed from the slip than with conventional slip casting; therefore, the "casts" were flabby.

(d) Contrary to prediction, marked difference in thickness of the deposit were experienced; this was ascribed to changes in potential over the article, probably arising from ionization in the water.

Regarding the prevention of faults attributable to electrolysis, de-ionized water and media other than water such as organic fluids have been suggested; the latter have proved successful but, except for special cases, seem to be prohibitively expensive.

Two processes, using zinc as the electrode metal, have been developed (Chronberg, 1978). The metal was oxidized by reactive atomic oxygen which became combined; the cause of trapped bubbles of oxygen leading to pin-holes was thus removed. Moreover, the zinc oxide formed retained its white colour and thus did not cause any dark specks in the ware.

The problem of flabby deposits could be lessened by using slips of very high solids concentration.

The uneven thickness of the deposit was remedied by ensuring that the gap between the two electrodes was constant (Ryan and others, 1981).

In an effort to fabricate complicated shapes by ED, Ryan and others (1981) developed what they called "cathodeless electrophoresis", reversing polarity during deposition, made possible by insulating the various parts of the mould from each other.

ED, as far as the author is aware, was first installed on a full industrial production scale around 1977 in France, using Chronberg's processes.

The possibility of coating a piece of pottery with glazes and colours, while still in the mould, is an exciting spin-off in the field of ED.

4.3. *Microwave Casting*

A British manufacturer of ceramic machinery, under licence from a U.S. company, has introduced microwave energy into the slip-filled mould for controlled and rapid drying of ware. The system is arranged for continuous operation on a rotating table. During the casting process microwaves pass through the mould and the slip, heat being generated in the centre of the article. When the mould is drained and tipped, heat is retained within the ware, resulting not only in faster cast formation, and hence increased production, but also in evaporation of water from the mould to enable it to be used again immediately. Moulds can thus be used up to twenty times a day, compared with the conventional two or three casts per day. This means a drastic reduction in mould requirements (Anon., 1986).

5. Shaping in the Dry State (Dust Pressing)

Dry pressing or semi-dry pressing (signifying a moisture content of up to 3 and 8 per cent respectively) is the most advanced of all pottery-forming methods, dispensing with two of the four elements, viz. *water* and *air*, essential for plastic and liquid shaping.

The idea of producing dinner plates by dry pressing is not new. Trials with this aim in mind were carried out in Spain in the nineteen-twenties; Austrian technologists investigated the problem in 1957 (Jung, 1983). Plates were first dry pressed on an industrial scale in the early nineteen-seventies (Ciarrapico, 1979).

Tiles have been dry pressed for many years, with metal dies which are filled with spray dried granules, and it seemed reasonable to assume that dinner plates could be made by the same process. With a simple shape like tiles, the pressure applied during the pressing operation is the same throughout; uniform density is, therefore, achieved. However, with curved articles, such as plates, ordinary dies are incapable of producing uniform compaction. Therefore, there are different densities at different points of the plate. This, in turn, leads to distortion during firing.

ITTP—D

With plastic shaping there is a "gliding" of particles over each other. This *"plastic flow"* capacity is responsible for a remarkably uniform density in the shaped piece. Dry particles do not glide, or at least are incapable of doing so sufficiently to even out differences in density, even with the aid of lubricants.

It was realized that equal pressure must be applied over the whole pressed piece if the deleterious effects of uneven compaction were to be avoided. Such a condition is met with *"isostatic"* dry pressing (IDP). The basic principle of isostatic (meaning "equal pressure") pressing is simple: uniform hydraulic pressure, applied via a liquid medium to a flexible tool (i.e. a bag) containing the powder to be pressed, causes this powder in the tool to be compacted uniformly.

However, the actual operation of IDP is somewhat complicated and involves a number of time-consuming operations, such as filling the dies, closing dies, fixing pressure head, actual pressing, stress relieving and various steps in the removal of the pressed article. A new process has been developed, combining IDP with a normal rigid die, thereby eliminating the numerous operations of "classical" IDP and at the same time retaining its great advantage of equal pressure (Schlegel, 1975). The face of the plate is formed by the rigid upper die, its back by the elastic membrane. The actual technique is like ordinary die pressing, with the lower die being in the form of a flexible tool, instead of another rigid die. An IDP machine, manufactured by a German engineering firm, is shown in Fig. 4.11.

Some German and Italian firms have developed techniques of pressing plates in the vertical position. This has the advantage of the granulate being profiled in the die cavity before pressing, thus allowing non-circular plates and steep-sided bowls to be pressed (Cubbon, 1982). However, the full isostatic effect is not realized with this arrangement; hence completely uniform density is not achieved. Vertical IDP, while satisfactory for porous bodies such as earthenware, is not always suitable for bodies with a high melt content and thus more liable to distort.

IDP has the following advantages over roller making:

1. IDP lends itself to fabricating shapes other than circular plates and bowls.
2. IDP does not require plaster moulds; dies and flexible membranes, although more expensive initially, outlast plaster moulds, thus bringing about a reduction in costs.
3. IDP requires less factory space (no mould storage needed).
4. IDP does not produce "scraps"—no wastage of body.
5. IDP does not involve drying—saving of fuel.
6. IDP is not liable to faults arising from drying, as no shrinkage occurs before firing.
7. IDP does not give faults like "centre pips", present in roller made plates, due to the orientation of particles, as evidenced by thickness

shrinkage in roller-made ware being almost twice that of diameter shrinkage, whereas with IDP diameter and thickness shrinkages are identical.

8. IDP allows greater size control and thus offers the designer greater scope as regards closely fitting patterns.

9. IDP lends itself to even more overall automation than does roller making.

FIG. 4.11.   Isostatic press for large tableware, type P1 550 G. By courtesy of Dorst Maschinen und Anlagenbau.

It has been claimed that changing from one shape to another on the IDP machine takes comparatively little time and the advantage of greater flexibility of the roller machine, as mentioned above, is therefore in doubt.

There used to be a definite disadvantage of IDP; it was incapable of producing the high-quality embossed and scalloped edges obtainable with roller making. This obstacle was removed, however, in the middle nineteen-eighties.

The most significant disadvantage of IDP is economical; a high capital outlay is required, not only for the pressing machines but also for the spray dryer which is essential for IDP, but not absolutely necessary, though desirable, for roller making.

There are reasons to believe that IDP will be applied to shapes other than plates. Experiments to isostatically dry press cups were carried out in the nineteen-sixties (Papen, 1967). There is no doubt that IDP is *the* shaping method of the future.

Further Reading

General: **Kenny (1954)**; Cubbon (1982, 1984); Singer and Singer (1963, pp. 716–54).
Plasticity: Moore (1965).
Roller making: Cubbon (1982).
Ram pressing: Cubbon (1982); Ott (1975).
Impact forming: Cubbon (1982); Bradshaw and Gater (1980).
Injection moulding: Handke (1974).
Slip casting: Scientific basis: Phelps (1982); Moore (1965); Practical aspects: Herrmann and Wich (1969).
Pressure casting: Olgun and others (1970).
Electrophoretic deposition: Ryan and others (1981); Hennicke and Hennicke (1982).
Dust pressing theory: Groenou (1982).
Isostatic pressing: Niffka (1979); Dube (1980); Zimmermann (1987).

5

Drying and Finishing

The *last* chapter ended with the statement that dry pressing was the shaping method of the future. As no water is involved, dry pressing does not require drying; this means greatly reduced manufacturing costs through savings in energy (and floor space), through fewer rejects and through removing a bottleneck in a fully automated production line. Yet *this* chapter is headed by the very subject the elimination of which is responsible for these benefits. However, at least for the time being, many articles, including all studio pottery, are still shaped with water as an essential ingredient. Doing away with the drying of such ware would cause it to be shattered in the early stages of the firing.

Apart from the water absorbed by the mould, the residual moisture in the piece can only be removed by *air*, the third of the four elements of which the world is made, according to the ancient Greeks who realized the importance of air in ceramic manufacture. The laws of Athens stipulated a drying period in the sun of five years for bricks to ensure dimensional stability (Amison and Holmes, 1985). Drying in air without applying heat is still practised by some individual peasant and studio potters today. It is the safest way, but, of course, too slow for industrially made pottery. Drying is accelerated by heat which, however, has to be kept to below a certain maximum depending on air humidity. In certain plants kilns for firing double as dryers in the initial stages, but this is far from ideal, if only because some of the finishing has to be done after drying, the two often being interwoven. Drying is interrupted so that the piece can be "finished", e.g. the handle fastened to the cup, etc.

The term "finishing" does not signify the final process in pottery, it merely means that the piece is made ready for firing. Even then it has still a long way to go before it is actually finished in the normal sense of the word: glazing, "glost" firing, decorating, followed by one or more decorating firings, have still to be done.

1. Drying

1.1. *Mechanism of Drying*

Drying in pottery can be defined as the removal of water from a granular material (the pottery body) by *evaporation*. This involves both the transfer of heat from the surrounding environment to the solid-water system and the simultaneous transfer of water vapour in the reverse direction.

Initially evaporation occurs only at the surface of the piece, the evaporated water being partially replaced by water that migrates from the interior to the surface. As drying progresses water can no longer move to the surface (for reasons indicated below) and is forced to evaporate within the pore system of the piece, the resulting vapour diffusing to the surface before it is removed.

In ceramic bodies the water reacts physically or electrochemically with the surface of the clay particles. When mixed with water, clay particles are surrounded by stable water films of appreciable thickness. The surface of the clay mineral crystal is negatively charged and attracts any positively charged ions in the water. Water whose molecules close to a clay particle are strongly attracted to its surface is referred to as "bound" water. Water evaporated from the surface of the plastic piece is replaced by "unbound" water migrating from the interior. Consequently the particles come closer together and the piece therefore shrinks. When the clay particles, each still surrounded by its film of bound water, come into contact with each other movement of water to the surface is impossible and shrinkage virtually stops. Further drying causes the water, held in the network of pores formed by the particles, to evaporate; this water is replaced by air which has started to penetrate the piece at the drying surface.

Another method of removing water, besides evaporation, is *dehumidification* (Amison and Holmes, 1985): water vapour is removed from the air in a drying chamber by condensing it on the cold surface of a heat pump. The much drier air then passes through the condenser section and gains some of the latent heat of vaporization, given up by the condensing refrigerant. The warm air thus obtained is then used to dry the ware before again returning to the evaporator section of the heat pump.

1.2. *Transfer of Heat*

Ceramic ware can be dried by *convection, radiation*, or *conduction*. Often the three processes are combined.

Drying by convection is the most widely used method in the pottery industry. A current of warm air is passed over the surface of the moist article. This airstream supplies some of the heat required to evaporate the water and removes the vapour from the surface, transporting it away from the drying system.

Infrared drying is a form of drying by radiation where the transfer of heat proceeds from a high-temperature source, usually a series of powerful bulbs, to a low-temperature source, the pieces, not the intervening air, being warmed. Drying by infrared has been claimed to produce uniform heat transfer and therefore to allow more rapid heating than is otherwise permissible in pottery drying. Infrared drying is used for fired ware after glazing where a high rate of heating is not harmful.

Transfer of heat by conduction can be defined as the flow of heat through a body from positions of higher temperature to positions of lower temperature. It is operative in the later stages of drying since the transfer of heat from an outside source by convection and by radiation to a solid is restricted essentially to the surface.

High-frequency drying has potential applications in the pottery industry (Copeland, 1952). Electrical insulators, clay included, under the influence of very high frequency, consume a certain amount of electrical energy and transform it into heat. The heating starts in the centre of the piece which is always hotter than the sides, contrary to all other forms of drying. This is of great advantage because stresses, causing cracking, especially in large and thick-walled pieces, are avoided. The frequency used is around 20 Mc/s. The method is not economic because of the present very high cost of generating high-frequency power.

Microwave drying, although efficient and fast, is not generally used because of high capital costs. However, in Japan (Amison and Holmes, 1985) microwave heating is used to cause rapid mould release of roller made cups. Vacuum is applied to assist evaporation (the effect being similar to the phenomenon of water boiling at a lower temperature at high altitude than at sea level).

1.3. *Types of Dryer*

Great strides have been made in the design of dryers. Gone are the days of the huge drying stoves equipped with shelves and heated by steam pipes and occupying up to two-thirds of the shop floors of the making shops. In modern factories the emphasis is on economy of space and increased efficiency. Most modern dryers are of the continuous type. They are equipped with heating units, fans, and controls of temperature and humidity. Often waste heat from the firing kilns is used for heating dryers but they can also be heated directly by steam or by gas, oil, or other types of fuel.

Tunnel dryers consist of long tunnels through which the moulds or unmoulded ware travel on trucks, on conveyor belts or on racks suspended from an overhead monorail.

Mangle dryers consist of shelves suspended on endless chains. The shelves carrying wooden boards filled with moulds or ware move vertically through a drying chamber provided with baffles. The moulds or unmoulded pieces travelling from a loading point up one side and down the other to an unloading station, usually pass from a moist, hot atmosphere to a dry, hot atmosphere.

Dobbins provide perhaps the most popular type of modern pottery drying. The ware moves in a horizontal circle about a central vertical axis, very much like a roundabout. There are usually about eight to twelve compartments, one for loading, six to ten in the completely enclosed drying zone, and one for unloading. Modern dobbins incorporate jet drying, jets of hot air being directed on to the ware. This causes even drying and enables the speed of drying to be increased. Dobbins require relatively little floor space and, together with the making machines, often form part of a unit.

Jet drying is not confined to dobbins. The amount of energy required for conventional drying is as high as 22 per cent of the total energy consumed in a pottery factory (Amison and Holmes, 1985), whereas with certain types of jet dryers it is only 10 per cent (Netsch, 1978).

One of the most modern dryers for flatware is the elevated jetting dryer (Caswell, 1977): the ware is conveyed vertically and then horizontally through the drying zone. This special construction requires very little floor area. Apart from the hot air jets, infrared heating is incorporated in this type of dryer. Its great advantage is speed of production which implies very few moulds being required.

The same criteria apply to an automatic cup-making unit, produced by a British engineering company; in the drying section of this plant air of 300°C is blown down into the moulds. The hot air comes into contact with the wet surface of the cup so that heat is quickly dissipated by evaporation and conduction, resulting in an even overall moisture content. The actual drying time is merely 80 seconds. The moulds show no sign of deterioration because of the fast dissipation of the hot air which ensures that the temperature of the plaster does not exceed 45°C at any moment. In fact the plaster moulds (being protected by the cup against over-heating) are said to last twice as long as under traditional conditions (Waye, 1973). Although subjected to a blast of high-velocity hot air, the temperature of the cup itself does not rise materially. At the very fast rate of evaporation, due to the combination of heat and turbulence, the adsorption of latent heat has a cooling effect on the surface from which evaporation is taking place. An analogy is the cooling effect of ether on the skin.

1.4. *The Process of Drying*

The sequence of drying is not the same with each method of making.

The studio pottery usually allows his freehand shaped or thrown pieces to dry naturally in air without placing them in a potter's stove or dryer. Long before the ware is completely dry he may apply an engobe and slip decoration (Chapter 8, p. 156) and only then finish the drying. The state at which he decorates the piece in this manner is termed "leatherhard"; this corresponds to the condition when the clay particles (still surrounded by films of "bound" water) have come into contact with each other.

Plates and other flatware made by jiggering are dried completely while on the mould in one of the types of dryers described. The pieces shrink as a result of the first stage of removal of water; ware and moulds are dried at the same time.

Hollow ware produced by jollying is usually dealt with in the same way as flatware, although cups are often dried to the leatherhard or somewhat drier state when handles are attached and the drying then completed. The moulds are dried separately after unmoulding of the ware in the leatherhard state.

Castware is first dried inside the moulds, either in cold air or in dryers, removed from the moulds when sufficiently hard, allowed to dry in air at room temperature until leatherhard, when parts produced singly in moulds are stuck together. This is followed by heating to complete dryness.

Drying is one of the most important processes in pottery production. A number of faults, warping and cracking being the most frequent ones, are caused by uneven drying. Very often these faults do not become apparent until after firing. Drying is governed by the following factors:

1. The plaster moulds.
2. The composition (and grain size distribution) of the body, particularly the types of clays used.
3. The shape, size, and thickness of the piece, especially in relation to the grain-size distribution.

When drying plaster moulds it is important that the temperature is not so high as to cause calcination and that the moulds are cooled sufficiently slowly to avoid cracking.

The following procedure has been recommended:

(a) If 5–20 per cent free moisture remains in the moulds a drying temperature of up to 105°C is permissible.
(b) Substantially dry moulds should not be subjected to more than 45°C.
(c) The temperature at the dryer exit should not exceed 37°C.
(d) Relative humidities above 10 per cent should be maintained throughout drying.
(e) Rapid circulation of air is more effective than high temperature (a very important point).
(f) Warm moulds should not be exposed to direct impingement of cool air.

Bodies made of very plastic clays or mixtures having a high clay content require particularly careful drying. Such bodies have a great number of colloidal particles. The path from the interior to the surface is therefore long and complicated, especially in large, thick-walled pieces, and the movement of water slow. It is therefore important that drying to the stage when shrinkage has ceased should be slow, as otherwise the smooth mechanism of drying cannot function and stresses are set up which lead to cracking and warping. These faults are basically due to differential shrinkage of the body which may be caused by different rates of loss of water from the surface and the interior, by uneven moisture distribution within the piece prior to drying, and by preferred orientation of the platelets of clay, which is practically unavoidable with plastic shaping or slip casting.

Even drying is particularly important with plates. If the rim dries in advance of the centre, so-called "humpers" result, the centre of the plate being arched up. If the centre dries in advance of the rim, so-called "whirlers" are produced, the centre dropping so that the plate ceases to rest on its foot (see Fig. 5.1). Whirlers can also be avoided by a convex surface (the so-called "spring") on the plaster mould.

When the piece has reached a stage beyond the leatherhard condition it can be dried quickly without danger of warping or cracking.

The distortion of clay (or clay bodies) during drying depends to a large extent on the way it has been previously treated. The potter is aware of the phenomenon known as the "memory" of clay (already referred to in Chapter 4, p. 59). For example, if a clay bar in the plastic state is mechanically deformed during shaping, it tends to revert rapidly to its original shape (elastic recovery) on removal of the deforming force. Interlocking of the particles, however, prevents complete relief of stress. Subsequently during drying, water moves within the bar and unlocks the particles. Under the influence of the residual stresses, these particles move in such a way that the bar continues to revert to its original shape. Clay thrown on the wheel without being subjected to further mechanical forces does not distort during drying.

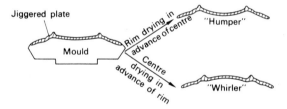

FIG. 5.1.   Drying distortion of jiggered ware. (After Ford (1964), p. 61.) Reproduced by permission of The Institute of Ceramics.

On the other hand a piece which has been subjected to markedly uneven stresses in shaping tends to "go out of shape" during drying. To guard against this a "ring" is placed on cups and other hollow ware (even on pieces that had been thrown but subsequently "jollied") as a support immediately shaping is finished. The ring is usually made of plaster, sometimes metal. "Ring" is a misnomer as the support is more of the shape of an open shallow cone with its top part cut off. The angle of the cone is governed by the time–shrinkage relationship. It remains with the piece while it is being dried in the mould and after unmoulding, i.e. until drying shrinkage is complete.

The drying operation has, for a long time, meant a blockage in integrated automatic pottery manufacture. The ultra-rapid dryers referred to above (section 1.3, p. 86) have not only removed this bottleneck but have also been instrumental in reducing faults, such as warping and cracking during firing, attributable to uneven drying.

2. Finishing

The stage of finishing is not always clear cut and some of the processes described below such as "sticking-up" might have been relegated to the previous chapter.

2.1. *Turning*

Turning is required for thrown ware to provide a pot with a foot. This is usually done the day after throwing when the pot has partially dried but not quite reached the leatherhard state. The pot is centred upside down on the wheelhead and fastened with wads of clay. The potter then removes excess clay and forms a shapely foot with a suitable tool while the wheel is spinning.

In industrial pottery turning is only applied to cups and other jollied hollow ware, such as sugar basins, made of bone china. The outside of the cup is turned to the required shape on a simple lathe:

(a) To remove surface roughness arising from the low plasticity of the bone china body. (A high surface finish is expected in bone china.)

(b) To avoid the complication of having to use a three-part mould where there is an undercut in the design, or of having to stick a separate foot on to the cup.

Turning, a skilful operation, used to be carried out freehand with the aid of a simple tool; but the process has now been largely mechanized.

The turning of bone china cups is done in the leatherhard or slightly harder state. The leatherhard is preferred to the dry state because the clay body is then somewhat elastic and considerably stronger than when dry (even then great care has to be taken to avoid breakage). Sticking handles to cups is also preferable in the leatherhard state for reasons to be explained below.

Turning has become a dying art as far as bone china is concerned. Roller-made bone china cups are of sufficiently high standard not to require turning.

2.2. *Cutting and Trimming*

These operations apply mainly to slip cast pieces. Before unmoulding or while the pieces are partly unmoulded the potter cuts away the spare formed by the riser in hollow ware or by the casting hole in the case of handles, spouts, or the individual parts making up figurines.

Since castware is very often made in multiple plaster-of-Paris moulds seams are unavoidable where the separate parts of the mould meet. On ware made with new moulds the seams are very fine, but they become more noticeable as the mould wears. These seams are removed with a sharp knife when the piece is fully dried. If done in the leatherhard state, the knife, while cutting away the spare, exerts some pressure on the body of the piece at the site of the seam, which becomes visible again after firing (another instance of the "memory" of clay). Seams are removed mechanically from cup handles in fully automated plants.

2.3. *Sticking-up*

As already indicated in Chapter 4, it is not usually practical to produce cups with handles or tea pots with handles and spouts in one operation. As with figurines, where there are undercuts, the individual pieces have to be made separately and fixed to each other afterwards.

The fixing medium or "stick-up slip", to give it its proper name, is simply ordinary clay body in the form of a very thick slurry having almost the consistency of a paste. It is prepared by mixing dry scraps (broken plates or other pieces in the dry state) with water to the right consistency. Recommendations have been made to make the stick-up slip stickier, for instance by adding a deflocculant and magnesium sulphate, but untreated stick-up slip is usually satisfactory.

Sticking-up can be done either in the leatherhard or in the dry state. In the dry state it requires less skill as the pieces cannot then be distorted. With dry ware the water from the fixing medium is immediately absorbed by the open pores of the ware; re-adsorption of water in a dried piece has a weakening effect apart from causing the stick-up slip to dry out, thus preventing it from performing its function. To avoid the absorption of water by the dry pieces from the stick-up slip, water is applied to the piece at the points of the joint prior to sticking-up.

A certain firmness of touch is essential with leatherhard pieces and sticking-up in this state is preferable as a better joint is produced. Here the difference in water content of stick-up slip and parts to be stuck up is much less. The water from the stick-up slip cannot easily be absorbed by the parts because in the leatherhard state there are no open pores; the clay particles are still surrounded by their film of "bound" water.

Where large surfaces have to be united the piece is roughened by criss-cross lines to provide a mechanical key.

Cups are often stuck up when a little drier than leatherhard. The handle is dipped in the stick-up slip and firmly pressed on to the bowl of the cup and some of the surface slip removed at the joint with a brush pencil. With high-quality ware, such as bone china, the cup is left for a certain time (while other cups are being prepared in the same way) before it is finally trimmed. By then some of the water from the stick-up may have evaporated or very slowly been absorbed by both bowl and handle so that moisture contents will have evened out. The hardened surplus stick-up material is then removed with a wooden spatula-like tool which also serves to consolidate the joint, firmly uniting the two parts.

The method of fixing handles or spouts to tea or coffee pots or for uniting different parts of a figurine is essentially the same as for cups. Regarding figurines, skill is required in making the joints appear invisible; some aftermodelling is necessary, especially with worn moulds.

The figure caster does all the finishing operations as well as the casting himself. With tableware there used to be a strict division of labour, e.g. for cups the work was divided between thrower, jollier, turner, handle caster, and "sticker-up".

In modern pottery factories the joining of handles to cups is automated within a conveyor belt system, incorporating cup making (usually by roller) and handle making (by impact forming, injection moulding, etc.). In any case, whether done by hand or mechanically, the joint should be stronger than the parts joined.

2.4. *Sponging, Fettling and Towing*

Before the ware is placed for firing, imperfections have to be removed; these include seams already referred to in section 2.2, p. 89, in respect of castware, including figurines. With cups, as well as castware, blemishes are often removed in the semi-dry state with water, using a sponge or brush pencil. This operation, known as "*sponging*", provides a smooth surface. Seams and other marks in castware are usually eliminated in the dry state with a knife-type tool, this process being referred to as "*fettling*". Fettling, followed by sponging, is also applied to plates and other flatware in order to remove the seams, produced where tool and mould meet, the plates being spun round on a rotating horizontal wheel head. In modern pottery factories this operation is fully automated, except for high-quality bone china plates for which tow is used to remove marks caused by mould imperfections; this is termed "*towing*". No water is used in this dry process, which is dusty and is, therefore, done under a dust extractor hood.

Further Reading

Theory and practice of drying: **Ford (1964, 1986)**; Scholz and Gardeik (1980); Singer and Singer (1963, p. 832).
Efficiency of drying: Amison and Holmes (1985).
Saving energy: Netsch (1978).
High-speed drying: Waye (1973); Caswell (1977).

6

Firing

Fire makes the potter's handiwork durable and more beautiful. In the end, of all the four elements which bring about pottery, earth is the only one that remains. However, the earth fashioned with water, dried by air, and hardened by fire is finally greatly changed. This change is permanent but the degree of change depends on the intensity of the heat treatment.

Prior to the introduction of scientific methods firing was the operation least understood and least under control in pottery manufacture. The experience gained in the hard school of failure, accident, and patient empirical experimentation was kept a close secret and it is therefore not surprising that very little is known of how our forebears tackled this most difficult and hazardous of all pottery problems.

In other industries the finishing of the product means the end of its processing. In pottery and ceramics generally, one is almost inclined to call "finishing" (as referred to in the previous chapter) the beginning because it is then that the most difficult operation starts.

What happens to the body during firing?

1. The Action of Heat on Pottery Bodies

1.1. *Physical Changes*

The physical phenomenon occurring when heat is applied to ceramics is known as "sintering". In this process particles which are in contact agglomerate, when heated to a suitable temperature, with a decrease in surface area and also in porosity of the aggregate.

Let us consider the sintering of a compact of particles of a pure, single-component substance, such as alumina. On heating, material flows from areas of low to high curvature, resulting in the formation of necks between particles. The driving force of this process is the resulting reduction in surface energy. The mechanism whereby material flows is still in doubt. It is considered, by some investigators, to be *plastic flow* due to surface tension, but the balance of evidence now favours an ion by ion diffusion process. In any case the result is that the distance between the centres of particles decreases, leading to shrinkage of the compact and a decrease in porosity. With further heat treatment (temperature and/or time) the porosity continues to decrease and the original multiparticle structure disappears. Eventually, all the interconnected porosity is removed and only a number of closed pores—some within individual crystals and others at grain boundaries—remain. Gases are trapped in these closed pores and exert pressure resisting further densification. If sintering is done in an atmosphere of hydrogen or oxygen, a completely non-porous transparent material (alumina) of theoretical density is obtained as these gases diffuse rapidly through alumina.

In the complex system which comprises pottery, sintering is complicated by the presence of a liquid phase due to the flux content of the body. The solid phases are wetted by this liquid and are soluble in it to an extent depending on temperature. Under these conditions the liquid penetrates between the solid particles and draws them together by pressure and capillary forces. If sufficient liquid is present all open pores are filled with liquid and a completely dense body is produced. Such a body is called "vitrified". Sintering to complete density in the presence of a liquid phase is known as "vitrification".

1.2. Chemical Changes

The chemical reactions of pottery bodies during firing are best studied by examining their microstructure. This consists of crystals, an amorphous mass (glassy matrix), and voids. The *optical microscope*, within limits, is a useful instrument for identifying crystal phases and for illustrating general microstructure; it is usual to view thin sections (not thicker than 30 μ) with transmitted light under crossed nicols. However, it does not reveal the *very fine* crystals present in pottery bodies. The optical microscope can therefore be misleading unless supplemented by *X-ray diffraction*. *Thermal expansion* also allows a reliable and easy check of the various phases present. For instance, cristobalite, due to kaolinite lattice break-up, too fine to be visible under the optical microscope, is shown up by a very sharp rise in expansion at its inversion temperature. *Differential thermal analysis* is also valuable for identification.

The most useful method, however, for studying the reactions that occur in pottery bodies during firing is *electron microscopy*. Here replicas of the body are examined. In view of the very high magnification possible (50,000 times and higher, compared with the maximum of 1000 times with the optical microscope) a true and very detailed picture of the microstructure is possible. The *scanning electron microscope*, due to its great depth of focus, produces photomicrographs of a three-dimensional character; furthermore, there is no need for preparing thin sections or polished specimens as with the optical microscope.

Electron-probe micro-analysis does not only provide a photomicrograph, but also an elemental distribution; it thus represents an ingenious combination of electron microscopy and chemical analysis.

1.2.1. *The "Classical" Triaxial System (Alkali-Alumina-Silica)*

The main constituents of pottery bodies are usually clay, feldspar, and quartz. In the lower temperature ranges of firing pottery the separate changes of these individual materials as described in Chapter 2 take place. The most important of these changes is the initial break-up of the kaolinite lattice. This means that in a large kiln, containing 20 tons of ware, 1 ton of water has to be evaporated and removed. The first interaction between the constituents occurs when feldspar starts to melt at about 1010–1100°C (depending on the alkali ratio). It then dissolves an appreciable amount of the quartz introduced in the body as well as the cristobalite developed as a result of the kaolinite lattice break-up. This reaction increases the viscosity of the melt. The increasing temperature, however, tends to counteract the effect of silica on the viscosity.

Simultaneously with this solution of quartz in the molten feldspar, small "scaly" crystals of mullite (*primary* mullite) are formed in the clay particles in the course of the kaolinite lattice break-up. This mullite formation is accelerated by alkali oxides present as an impurity in the clay.

If these reactions are arrested at, say 1100–1150°C, the liquid phase is either too small in amount or too viscous to fill all the pores. It does, however, cement the silica crystals (and the small scaly crystals of mullite) together, forming a strong bond. The type of pottery thus formed is *earthenware* (strictly speaking, feldspathic or English-type earthenware), individual silica crystals being embedded in an amorphous matrix.

If the reactions are allowed to proceed to 1200–1250°C sufficient liquid phase is present to penetrate practically all the interparticle interstices so that the body becomes fully dense. Such pottery is known as "vitrified hotelware" in Britain, as "vitreous china" in the United States, where it represents the main type of porcelain produced. Its microstructure does not materially differ from that of earthenware.

When the temperature is further increased, say to 1280°C, needle-shaped crystals of mullite (*secondary* mullite, much larger than the scaly crystals derived from the clay) grow out of the melt (Fig. 6.1A). The growth of these crystals causes the glassy phase to become less viscous. This more reactive glassy phase progressively dissolves the free silica crystals until they have all disappeared. Clusters of needle-shaped mullite crystals are then the only crystals visible under the optical microscope, as shown in Fig. 6.1C. The type of pottery represented by such a microstructure is "hard porcelain". It should be pointed out that the structure shown in Fig. 6.1C represents an ideal case, viz. high thermal shock-resistant laboratory porcelain with its initially low feldspar and quartz contents; normal tableware hard porcelain still contains some undissolved silica crystals (Fig. 6.1D).

With further increases in temperature in bodies containing relatively high amounts of feldspar, the more fluid melt formed attacks and redissolves the mullite needles (Fig. 6.1B) it helped to form initially. However, the scaly

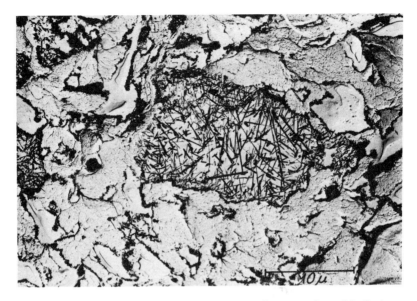

FIG. 6.1A. Microstructure as revealed by electron microscope of porcelain fired at cone 8 (1280°C) showing mullite needles crystallized from the melt. Reproduced by permission of K. Schüller (1964) and The Institute of Ceramics.

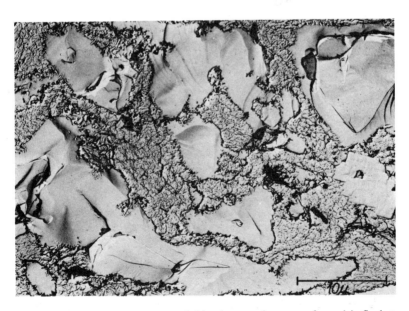

FIG. 6.1B. Microstructure as revealed by electron microscope of porcelain fired at cone 14 (1410°C) showing aggregates of primary mullite only, most of the secondary mullite needles redissolved. Reproduced by permission of K. Schüller (1964) and The Institute of Ceramics.

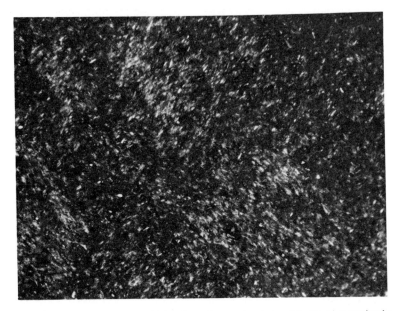

FIG. 6.1C.   Microstructure of high clay substance hard porcelain showing randomly orientated secondary mullite needles as the only crystal phase; crossed nicols; × 120.

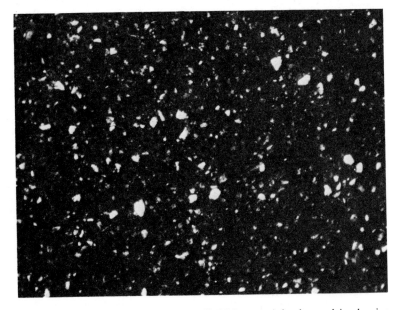

FIG. 6.1D.   Microstructures of "normal" (high quartz) hard porcelain showing undissolved or partly dissolved quartz as the only crystal phase; crossed nicols; × 120.

mullite due to the clay is not attacked unless there is a high amount of quartz in the body. In this case the aggregates of scaly mullite recrystallize to large needles and as such are again prone to redissolution.

As already indicated (Chapter 2, p. 14), china clays usually contain appreciable amounts of alkali. These are not always present in the form of feldspar but very often as mica. Mica can have twice the vitrifying effect of feldspar in bodies (Inzigneri and Peco, 1964).

The potash/soda ratio of the feldspar introduced into the body has a certain influence on the reactability; however, this is not as far-reaching as might be inferred from the behaviour of the two types of feldspar on their own (see Chapter 2, section 2.2.1, p. 29). It was shown that the speed of dissolution of quartz in soda feldspar melts compensated for the low viscosity of such melts; therefore, at high temperatures, the speed of deformation in porcelain with different feldspars did not differ significantly (Schüller and Jäger, 1979).

1.2.2. *The System Containing Calcium Phosphate*

This system concerns bone china. Bone consists of hydroxy apatite $(Ca(OH)_2.3Ca_3(PO_4)_2)$ which, on heating, starts to decompose at about 1000°C (St. Pierre, 1955) into beta-tricalcium phosphate $(Ca_3(PO_4)_2)$, lime (CaO) and water (H_2O). The following three processes seem to be involved when the bone china body is heated:

1. Some CaO from the bone combines with the china clay to produce anorthite $(CaO.Al_2O_3.2SiO_2)$.
2. All the remaining CaO from the bone goes to make beta-tricalcium phosphate $(Ca_3(PO_4)_2)$.
3. The remaining P_2O_5 combines with the other materials to form a glass.

Regarding the first body reaction, it was once thought that only the calcium in the $Ca(OH)_2$ was available for forming anorthite. However, Beech (1959) convincingly demonstrated that all the lime in the bone ash was available for anorthite formation. The total alumina from the china clay was converted to anorthite; the amount of lime derived from $Ca(OH)_2$ only would not have been sufficient for this reaction. Therefore, a portion of the lime must have been taken from $Ca_3(PO_4)_2$; this, in turn, implied that some P_2O_5 was liberated, combining with the constituents (not taken up in the main reaction) to form a glass. The phases present in the fired bone china body are thus:

	%
Anorthite	25
Beta-tricalcium phosphate	45
Glass	30

The (calculated) composition of the glassy phase is thus:

	%
SiO_2	65
Al_2O_3	15
P_2O_5	5
K_2O	15

It has been suggested that the nature and amount of glass in a bone china body possibly determines its physical properties to a greater extent than do the crystalline phases (St. Pierre/Rigby, 1956).

A body consisting of preconstituted material, viz. 45 per cent betatri-calcium phosphate, 25 per cent anorthite and 30 per cent glass (of the composition as calculated), fired to normal bone china biscuit temperature, gave a much more homogeneous structure than did the normal bone china body, consisting of the same phases, but prepared from conventional materials (Dinsdale, 1967). Moreover, the preconstituted body showed superior strength and translucency.

It was postulated from phase diagram studies that the bone china body composition was near a eutectic between tricalcium phosphate, anorthite and silica (St. Pierre, 1954). Reactions in mixtures, close to a eutectic, invariably proceed very rapidly, unlike those in the constituents of the classical triaxial system which make up all other traditional whiteware bodies; reactions in these are slow and extend over a wide range of temperatures, as will be seen from Fig. 6.2 which shows that the hard porcelain body vitrifies 200K below its final firing temperature, whereas the bone china body has still an apparent porosity (open pores) of 20 per cent at the vitrification temperature of hard porcelain, viz. only 30K below the final firing temperature of the bone china body.

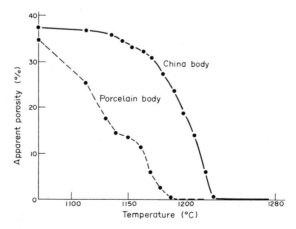

FIG. 6.2.   The influence of heat on the apparent porosity of bone china and hard porcelain.

1.2.3. Miscellaneous Systems

Among the combinations of alkaline earths with alumino-silicates only the one with *magnesia* (MgO) is of practical interest to the potter. It forms *cordierite* ($2MgO.2Al_2O_3.5SiO_2$) which, because of its low thermal expansion and hence high thermal shock resistance, is eminently suitable for cooking ware (see Chapter 9, section 2.3, p. 193). It is obtained by firing a mixture of the correct proportions of talc ($3MgO.4SiO_2.H_2O$), alumina (Al_2O_3) and china clay ($Al_2O_3.2SiO_2.2H_2O$) to give approx. $2MgO.2Al_2O_3.5SiO_2$. The true cordierite formula is nearer to a eutectic even than bone china, and has an impractically narrow vitrification range. Its composition is, therefore, modified to some extent; the alkaline "impurities", invariably present in clays, help towards this end.

Interesting bodies have been produced with *baria* (BaO), based on *celsian* (barium feldspar—$BaO.Al_2O_3.2SiO_2$) (Herrmann, 1964). Barium, introduced in the form of the sulphate, is the main ingredient of Wedgwood's *"Jasper"*; nothing seems to have been published on its microstructure.

1.2.4. Factors Influencing Chemical Reactions

One of the main characteristics of pottery bodies is the fact that, in firing, chemical reactions are not allowed to proceed to completion. The reactions are arrested before the system is in a state of equilibrium. Therefore, phase diagrams appertaining to pottery compositions are of very limited use. They bear no relationship to normal pottery firing practice, and besides, do not take into account the effect of certain factors governing reactions during firing, i.e. impurities invariably present in raw materials available, particle size, and firing atmosphere.

The finer the particles the more points of contact and the smaller the diffusion distance and hence the greater the reactivity. (To some extent fineness has a greater influence on reactivity than has composition.)

Clays usually contain alkalis, whereas the minerals introduced for their alkali content, viz. feldspathic materials, often contain excess alumina and silica. Such "impurities" do not matter as long as they are allowed for in the compounding of the body. Alkaline earths usually present in small amounts in clays and feldspars act as fluxes. The most important impurity, especially in clays, is iron oxide. "Common" pottery which sometimes consists almost entirely of red clay (Chapter 2, p. 19) contains up to 10 per cent ferric oxide (Fe_2O_3) and even more. This reacts with silica and forms a liquid, especially in the presence of alkalis and alkaline earths, at relatively low temperatures (approximately 1000°C). Therefore the firing temperature to produce a durable finished product need only be very slightly higher. Iron oxide is thus a vital constituent of this type of body. Titanium dioxide is another impurity normally present in clays. It appears to act as a catalyst (or mineralizer as it is often called in ceramic technology) promoting chemical reactions. Although white it greatly enhances the colouring effect of other metal oxides, particularly that of the iron oxides.

The effect of iron oxide is greatly influenced by the nature of the firing atmosphere, i.e. whether it be oxidizing or reducing. In a reducing atmosphere the ferric oxide is reduced to ferrous oxide which, with silica, forms liquids at much lower temperatures than those required to cause ferric oxide and silica to melt. Stoneware which contains a considerable amount of iron oxide (up to 8 per cent) is sometimes fired in a reducing atmosphere (especially where certain artistic effects are desired). If so, more liquid is produced than in an oxidizing atmosphere so that lower temperatures are sufficient for making the ware dense. The colour of such pottery ranges from dark purple to grey instead of red, brown, or yellow obtained under oxidizing conditions. The only other type of pottery fired reducingly during a vital part in the heating up is hard porcelain. Here the iron content is less than 0.5 per cent but even this small amount is sufficient to impart a pronounced greyish yellow tint to the ware if fired in an oxidizing atmosphere. The ferrous oxide formed as a result of the reducing atmosphere gives the porcelain a slightly bluish tint which many connoisseurs claim is more attractive than the pure white of bone china. (Formerly, Fe_3O_4 was thought to be produced in reducing conditions but Berg (1963) proved ferrous oxide to be the only stable form.)

The reducing gas used for firing hard porcelain is mostly CO (carbon monoxide). Other gases have marked effects on bodies generally. Water vapour reduces the viscosity of the glassy component and hence increases the vitrification rate; it also accelerates the conversion of quartz to cristobalite, thus increasing the thermal expansion. Fluorine, derived from china clay and some fluxes, can be found in kiln atmospheres; it can reduce vitrification and contraction, resulting in surface porosity and distortion of ware (Holmes, 1973).

A decisive factor, determining chemical reactions in pottery bodies, is speed of firing. Since the energy crisis in the early nineteen-seventies there has been an increased awareness of the need for saving fuel. One effective way of conserving energy is to shorten the firing time. This implies that the actual heatwork is reduced, the chemical reactions having insufficient time to procede. Firing to a higher temperature compensates for the loss of heatwork. Earthenware cups could be fired satisfactorily in 1 hour (as compared with the normal 24 hours) simply by increasing the usual firing temperature by 70 to 100K. A similar treatment with bone china cups did not work, since the desired translucency could not be achieved (Bull, 1982). Increasing the firing temperature is, to a large extent, self-defeating in an attempt to save energy. Adjustments to bodies is a more effective means of compensating for the loss of heatwork as a result of fast firing.

The obvious way is to introduce more flux into the body. This can be done by increasing the existing alkali content, by incorporating special glass fluxes, but most expediently by small additions of alkaline earths, mainly lime (CaO).

Making the crystalline silica more reactive by introducinmg it in a more finely divided state is another, even more effective, way of adjusting the body. With fast firing there is insufficient time for the quartz grains of conventional size to be taken into solution by the feldspathic melt. Large, undissolved free silica crystals have a detrimental effect on the thermal shock resistance of hard porcelain. Much finer quartz has counteracted this drawback. Holmström (1980) obtained equal and even superior physical properties in porcelain fired to 1400°C and cooled within 3 hours by introducing quartz of a grain size of 10 to 30 μm, compared with a body fired and cooled in the conventional period and containing much coarser quartz. Moertel (1977) used quartz consisting of particles 96 per cent finer than 10 μ which were completely dissolved relatively early in the firing, thus producing a silica-rich melt which, because of its high viscosity, prevented distortion of the articles made. The maximum temperature was reached in 70 minutes and held for about another 70 minutes; cooling was rapid, amounting to quenching. The body contained mullite in the form of *spherical aggregates*, derived entirely from the break-up of the kaolinite lattice and was highly translucent. More slowly cooled specimens showed less translucency which was due to the crystallization of *needle-shaped* mullite from the feldspar melt (see Chapter 10, section 5.2, p. 210).

In a number of cases, contrary to expectation, there appeared to be no need for altering body composition for fast firing.

2. The Placing of Ware

2.1. *Supports*

Throughout the manufacture of pottery precautions have to be taken against the manifestations of the "memory of clay". This is even more important in firing than in drying. Very often this "memory" does not reveal itself until the vitrified state. Broadly speaking, it is the more pronounced the higher the clay content and the more plastic the clay used.

When the piece shrinks there is movement within it and hence a potential danger of distortion. If the clay were perfectly homogeneous initially and if no appreciable stresses were introduced into the piece during making (as with throwing), then the risk of distortion is minimized.

Broadly speaking, stresses are almost invariably present in pottery made by industrial methods, and to prevent distortion the same expedients are applied during firing as during drying. Firing shrinkages (10–12 per cent) are higher than drying shrinkages (5–6 per cent) and the need for support during firing correspondingly greater. It is surprising that, while the complicated reactions described above proceed, causing softening, shrinkage, and movement, pieces retain their shape at all. Prevention of distortion is only possible by attention to detail in the making and drying processes and especially to the choice of supports during firing.

The simplest form of support for hollow ware is "boxing". For example the rims of two cups are placed together so that they keep each other straight. If shape does not allow this type of support plaques are used. Usually these are of the same body so that the support shrinks uniformly with the piece, in which case the plaque can only be used once. In some instances a refractory plaque is used which does not shrink and is designed so that the piece, when inverted over the plaque, "rides" on the plaque during shrinkage. Such refractory plaques can, of course, be used many times. The materials used for refractory supports are similar to those for kiln furniture, describe below (see section 2.3).

Figures require elaborate supports, usually in the form of "props" made of the same body.

Plates can be stacked, provided that they are not glazed, and are kept straight by layers of sand or alumina powder between them.

2.2. *Kiln Furniture*

With the exception of unglazed plates and saucers, pottery cannot be stacked during firing because of its shape and the movement which occurs as a result of shrinkage. Therefore special refractory structures known as "kiln furniture" are used, on or in which ware can be placed for firing. Dodd (1967) restricts the term to "pieces of refractory material used for the support of pottery-ware during kiln firing". In the present context kiln furniture embraces all refractory structures, including shelves, etc., used in kilns (cf. White, 1953).

The most common kiln furniture for firing pottery used to be refractory containers known as "saggars". They usually consist of 50 per cent fireclay and 50 per cent "grog" (crushed and graded particles of broken fired saggars). The purpose of grog is to prevent cracking and distortion of the saggars during firing.

When pottery was fired with solid fuel (coal or wood) saggars offered a convenient protection against flying ash and direct impingement of the flame. However, saggars took up valuable space quite apart from requiring extra heat. Therefore, and bearing in mind that no protection of the ware is needed with modern fuels, saggars have largely been replaced by refractory slabs called "bats" on which ware is "open placed". The bats are supported by refractory "props" and, as much more ware can be placed within a given space, compared with saggars, the method of open placing is much more economical. However, saggars are still the most suitable kiln furniture for firing plates and saucers because of their round shape. Saggars for firing plates and saucers are known as "setters" (Fig. 6.3).

In some modern factories the following steps of manufacture are fully automated (Maget, 1979; Reichl, 1979).

Placing unfired ware in saggars or setters.
Stacking setters into bungs.
Placing bungs on kiln trucks.
Transporting bungs to unloading station after firing.
Removing fired ware from saggars and setters.
Polishing, i.e. foot polishing of plates.
Returning saggars and setters to starting point.

In the most sophisticated plants the placing process is computerized. Different individual programmes (concerning various shapes and sizes of ware) can be fed into the control unit and memory bank by a programmer (Höltje, 1979).

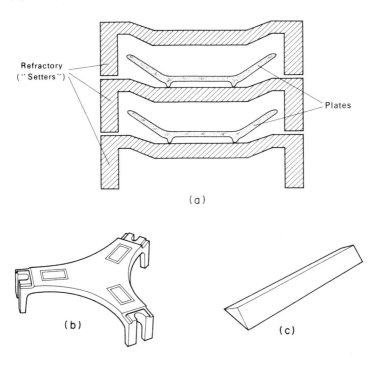

FIG. 6.3.   (a) Bung of setters for firing glazed hard porcelain plates. (b) Crank for firing glazed bone china plates. (c) Pin to fit in one of the three slots of the crank for the plate to rest on.

2.3. *Construction Materials for Kiln Furniture*

Pieces of kiln furniture have to be heated up from cold to the maximum firing temperature and then cooled down again to room temperature many times. These constant temperature changes imply that they are exposed to *thermal shock*. The 50/50 fireclay/grog mixture formerly used for saggars showed poor resistance to thermal shock. It was used because it was cheap, but kiln furniture made from it cracked during firing and had to be discarded after only a few times in the kiln. It was realized that it would be more economical to use kiln furniture made of more thermal shock-resistant materials. These are used mostly in precalcined form, i.e. as "grog" with only small amounts of high-quality refractory clays and/or temporary—organic—binders. The required shapes are manufactured by pressing, followed by firing to temperatures higher than those of their eventual use. Setters, bats, props and other types are now expected to last several hundred firings.

2.3.1. *Alumino-silicates*

Natural kyanite $(Al_2O_3.SiO_2)$ and synthetic mullite $(3Al_2O_3.2SiO_2)$ are more suitable as kiln furniture materials than are fireclay based products because of their higher refractoriness and greater resistance to thermal shock.

2.3.2. *Alumina*

Alumina (Al_2O_3) in the fused state is highly refractory but not so resistant to thermal shock as most of the alumino-silicates. It is expensive on account of the high temperature (2000°C) required for fusion. Bonding the alumina grains with a suitable flux reduces the high temperature of fusion and hence the cost, and also increases the strength of kiln furniture made from alumina thus bonded.

2.3.3. *Mullite/Cordierite Combinations*

Mixtures of mullite and cordierite $(2MgO.2Al_2O_3.5SiO_2)$ have proved the most successful kiln furniture material for most pottery products. Cordierite, with its very low thermal expansion and hence outstanding thermal shock resistance, is not used on its own because of difficulties in manufacture (as already hinted at above under section 1.2.3). Suppliers of refractories merely add talc $(3MgO.4SiO_2.H_2O)$ to mullite before fabricating the articles of kiln furniture. During subsequent firing cordierite is formed.

Further reactions seem to take place while the pieces of kiln furniture are actually in use; more cordierite is apparently formed and, therefore, the thermal expansion is further reduced. This would explain why an old piece of mullite/cordierite kiln furniture stands up to severe thermal shock, whereas a new article of the same initial composition is found to fail under the same conditions (Lovatt, 1984).

The maximum temperature at which mullite/cordierite kiln furniture can safely be used is 1300°C; above that temperature other magnesia compounds are formed; these do not only have high thermal expansions, but are also liable to melt and thus cause deformation of the kiln furniture. Mullite/cordierite cannot, therefore, be used for the glazing fire of hard porcelain.

2.3.4. Silicon Carbide

Silicon carbide (SiC) has the highest resistance to thermal shock of all kiln furniture materials. This is not only due to its low thermal expansion, but also (even more so) to its high thermal conductivity. It is ideal for hard porcelain. It is no longer used for temperatures below 1300°C, because it is more expensive than mullite/cordierite mixtures and it tends to oxidize in the prevailing—oxidizing—firing conditions in which general pottery, except hard porcelain and certain types of stoneware, is fired; this caused silica to be formed which manifests itself in a "swelling", i.e. an increase in the size of the piece of kiln furniture. Oxidation of silicon carbide can be prevented, at least to some extent, by coating with a layer of alumina.

The majority of types of silicon carbide for kiln furniture contain up to 30 per cent clay, acting as a binder. The lower the binder content, the longer the pieces last, viz. at 1400°C, 500 firings with 80 per cent silicon carbide, 1000 firings with 92 per cent silicon carbide (Harms, 1979). Materials with higher thermal conductivity have been developed (which guarantee even longer lives), i.e. "reaction sintered" silicon carbide (Klein, 1979) and silicon nitride (Lovatt, 1984), as well as combinations of silicon nitride and silicon carbide (Lovatt, 1984). Reaction sintering merely means exposing silica sand to atmospheres of carbon monoxide (CO) and nitrogen (N) respectively, during firing.

2.3.5. Zircon

This material (zirconium silicate, $ZrO_2.2SiO_2$) is used when chemical inertness to glaze vapours is required.

3. Types of Fuel

Practically any type of solid, liquid or gaseous fuel or electricity can be used for firing pottery.

3.1. Solid Fuels

Fine brush, dried grasses or twigs of wood were and still are among the solid fuels used by the highly skilled so-called "primitive" potters. Dried dung, shaped into bricks, has been a favourite source of fuel, as it burns rapidly and evenly, holding its shape as an ember which protects the pottery ware from too rapid cooling. Artificial dung can be produced by combining sawdust, flour and bentonite. This mixture gives black ware if added at the end of the firing to prevent access of air, thus creating a strongly reducing atmosphere which in turn is responsible for the black colour.

Wood, charcoal and coal are still used by studio potters today to obtain certain artistic effects, difficult or impossible to achieve by other means.

In the late eighteenth century wood was replaced by coal, which dominated pottery firing to nearly the middle of this century. Coal has now been ousted by gas or gasified liquid fuels and electricity. Compared with solid fuels these are clean and easily manipulated.

3.2. *Gaseous Fuels*

Some of the larger pottery works, particularly in Germany, used to generate their own *producer gas*, but because of lack in consistency (due to variations in the coal used), producer gas has largely been replaced by other gaseous fuels. The most widely used fuel for firing in Britain is *natural gas*, obtained from the North Sea. France and Holland have their own—land based—natural gas.

Town gas, used in countries without natural gas, contains a certain amount of sulphur which is harmful in the firing of pottery, giving rise to glaze faults, such as pin-holes. A special type of town gas, known as "Lurgi" gas, is practically free from sulphur, as are natural gas and other gaseous fuels, like *kerosene* and the two *liquid petroleum gases*, *butane* and *propane*. The latter types, requiring special installations on site, are used mainly in locations without piped gas supplies.

3.3. *Oil*

Oil is not generally used for firing pottery, chiefly because of its high impurity content. However, very light refined oils have proved successful. (A porcelain factory at Lille is reported to have been designed for oil burning as early as 1785; see Charles, 1964.)

3.4. *Electricity*

Heating elements made from strips or wires of nickel-chrome are used to about 1100°C. For higher temperatures elements of *Kanthal*, a Swedish product containing aluminium, cobalt, and iron, besides nickel and chrome, are chosen. Rod-shaped elements made of silicon carbide are only seldom employed in pottery firing because of their high cost.

3.5. *Comparison of the Various Fuels*

The right choice of fuel in a particular application is dependent on many factors.

One important consideration is, of course, price per unit of heat. On this basis coal would be the best and electricity the worst. Price per unit of heat is, however, not the only criterion. Availability is a prerequisite and for this reason different fuels are chosen in different localities. Cost of installation and maintenance of equipment is another significant point to be borne in mind. For instance, North Sea gas being readily available in most places in Britain has a considerable advantage in this respect over liquid petroleum gases, which involve expensive installations. The standard of the ware produced also influences the choice of fuel; oil, being impure, is obviously unsuitable for firing fine bone china.

Electricity, considered the cleanest and most easily manipulated type of fuel, is preferable for low temperatures, e.g. the firing of decoration on the glaze. With electric firing the kiln atmosphere consists of relatively still air which is more suitable for glaze and decorating firing than turbulent combustion gases. For higher temperatures, gaseous fuels are more economical. They are also more expedient where reducing atmospheres are required.

4. Types of Kilns

Pottery can be fired without kilns, indeed without any fixed structure. Tanzanian women potters still fire their beautiful pots in bonfires, following a complex and highly developed procedure and achieving exceptional standards of artistic excellence.

Modern, industrially made, pottery is fired in batches in *intermittent* (periodic) kilns, or continuously in *tunnel kilns*.

4.1. *Intermittent Kilns*

With intermittent or *periodic* kilns the ware is placed in the kilns when they are cold and is then heated up to a predetermined time-temperature schedule until the maximum temperature is reached. This temperature is usually held for a certain period (up to 3 hours) called the "soak". After the soak the heat source is shut off and the ware allowed to cool naturally in the kiln until it is cold enough to be emptied (Fig. 6.4).

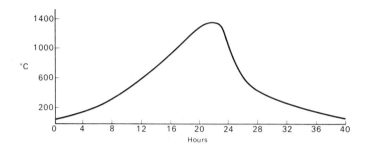

FIG. 6.4. Firing curve. The curve shows the actual heat treatment the ware receives. This applies to both intermittent and tunnel kilns. If in the case of a tunnel kiln the "hours" are replaced by the appropriate length measurements the firing temperature at each point in the tunnel is obtained.

4.1.1. Bottle Ovens

Although industrial bottle ovens are practically extinct—very few still exist as museum pieces—they are mentioned here as they represent the culmination of the art of building structures for firing pottery by solid fuel, from the early pits to the sophisticated treble-decker kilns for the heat treatment of hard porcelain (see Chapter 9, section 2.2.3, p. 188).

Bottle ovens were so called because their external shapes were those of a bottle. Saggars containing the ware were stacked in "bungs" (Fig. 6.3). When the oven was full the entrance was bricked up and the fires were lit in the fire mouths situated outside the kiln. The flames and gases passed up and around the bungs of saggars before finally leaving the ovens through the chimney. This was referred to as "up-draught". In some cases the flames were first directed to the vault of the oven. They were then forced down the bungs of saggars and through holes in the floor of the oven which lead through flues to the chimney. This system was known as "down-draught" and was claimed to produce better heat distribution. It was particularly suitable for reducing conditions and thus used for firing hard porcelain (see Chapter 9, section 2.2.3, p. 187).

The firing of bottle ovens is an art and the old fireman was a highly skilled worker. He had no aids to firing, except dampers. The conduct of the firing was greatly influenced by the prevailing winds and weather. The relatively even distribution of heat and accurate control of temperature achieved (to within 20°C in the case of bone china) seems a miracle to the modern technologist. Around 1980, in conjunction with the Gladstone Pottery Museum in Longton, Stoke-on-Trent, a bottle oven was resurrected after having stood idle for many years. The one remaining fireman with a few helpers lit it for the last time. It was a great historical occasion.

The firing of bottle ovens was an unpleasant and dirty job. Bottle ovens were exceedingly inefficient: only 3 per cent of the heat input was utilized for heating the ware; the remaining 97 per cent was lost through the chimney and wasted in the heating of the kiln walls and saggars. In Britain, bottle ovens were finally ousted by the Clean Air Act which forced pottery manufacturers to abandon smoke-producing kilns. Similarly, bottle ovens were discarded in other countries. As a result, townscapes of pottery centres which had been shaped and dominated by bottle ovens were completely transformed, none more so than Stoke-on-Trent. Before World War 2 this pottery city was permanently shrouded in a dense smoke haze which even the sun could not penetrate; by the early nineteen-fifties no trace of smoke was left.

4.1.2. Muffle Kilns

In the old bottle ovens the ware had to be protected from flying ash (derived from coal) and from direct impingement of the flames containing sulphur and other substances harmful to pottery. This protection was afforded by saggars, which implied a loss of useful space. In order to avoid this, especially with decorating firings, kilns were built with a refractory inner shell known as "muffle", within which the ware was placed to prevent contact with open flames. These interior structures were still used, to a small extent, after the change-over from coal to town gas. The availability of the much purer gaseous fuels have made muffles more or less obsolete. However, they allow operating with the cheapest or most readily available fuel. This could compensate for the higher cost of muffle firing arising from the need to heat the protective shell.

Muffles can also be incorporated in tunnel kilns.

4.1.3. Truck Kilns

In small potteries of the studio type the kiln may be box-like and self-contained so that the ware can be placed in it from the top or from the front, in which case hinged doors are provided. In industrial undertakings truck kilns consist of a shell and one or two kiln *cars* or *trucks*. These are furnished with refractory shelves on which the ware is placed. When full, the truck, usually running on rails, is pushed into the kiln shell.

In some cases the trucks are stationary and the kiln moves on rails. These types of kilns are referred to as *shuttle kilns* or *moving hoods* (see Fig. 6.5). In the moving-truck type the kiln shell is furnished with hinged doors but sometimes a permanent end wall is fixed to the moving truck, this end wall acting as a door for the kiln.

4.1.4. Top-hat Kilns

The top-hat kiln is a development of the stationary truck type. Instead of the kiln being moved on rails over the truck, the top hat (kiln shell) making

FIG. 6.5.   BIF moving hood china biscuit kiln, reducing firing time from 70 hours to 20 hours. By courtesy of Royal Worcester Spode Ltd.

up the walls and roof is lowered from a hoist over the stationary truck. This has two main advantages:

(a) There is no jarring or vibration when the kiln is moved, so that fragile ware suffers no damage.

(b) Cooling can be done very much more quickly and more evenly than with truck-type kilns by just raising the top hat a little immediately after firing. Opening the door in a truck-type kiln would cause very uneven cooling, leading to thermal shock cracking of the refractories, possibly of the ware itself.

As with the ordinary truck kilns, the top-hat kilns can be fired with gaseous fuels as well as by electricity.

4.2. *Tunnel Kilns*

The first *tunnel* or *continuous kiln* was built as long ago as 1751 (Hellot, 1959). Industrial tunnel kilns started to be used in the nineteen-twenties for low temperatures, viz. decorating firings (750–800°C). It was only since World War 2 that they were more widely introduced for biscuit and glazing ("*glost*") firings, electricity, but more often gaseous and liquid fuels, serving as heat source.

Tunnel kilns, as well as intermittent kilns, are usually built on site. A British firm has developed a *"package"* tunnel kiln made up of sections, called *modules*. These consist of prefabricated parts, constructed at the kiln builder's premises; they are shipped to the pottery factory where they are assembled (Barsby, 1971). Such package kilns, which are of the modern light-weight type, are ideal for locations without skilled labour for building kilns.

4.2.1. Conventional High Thermal Mass Tunnels

In such kilns the ware is placed on trucks which run on rails through the tunnel. Most tunnel kilns are straight but they can be circular; in that case there is an open section between the exit and entry ends where the ware which has been fired is emptied from the trucks, making room for fresh ware to be placed. Circular kilns are normally only used for the low-temperature firings required for decorations.

The temperature at any open point of the tunnel kiln remains constant (Fig. 6.4). Heat is therefore not wasted in repeated heating of the kiln structure as in intermittent kilns. Considerable quantities of heat are, however, used in heating the kiln cars each time they pass through the kiln. Improvements in tunnel kiln designs are aimed at reducing the mass of the kiln cars and the supporting structure. This also allows faster throughput of the ware.

4.2.2. Gottignies Multi-passage Kilns

In an attempt to save fuel, double tunnels were erected in which the trucks travelled in opposite directions. The heat given off the trucks in the cooling zone of one tunnel heated up the ware in the other. This idea led to the electrically heated Gottignies multi-passage kilns in which the ware is pushed through 16, 24 or 32 heated passages of small cross sections (Singer and Singer, 1963, p. 1043). The directions of travel are opposite in adjacent tunnels. At lower temperatures, i.e. for decorating firings, metal belts are used; at higher temperatures, refractory "bats" (slabs) serve as pusher mechanism.

4.2.3. The Trent Kiln

The slabs used in multi-passage kilns tended to crack. A British company developed a new type of tunnel, the so-called *Trent Kiln*, with the object of avoiding this defect. As before, the ware was carried forward on refractory bats; these rested on bricks which were pushed along in a refractory channel by hydraulic rams. The gaseous fuel required for the Trent kiln was claimed to be half of that needed for truck-type tunnels for the same amount of ware.

4.2.4. Roller Hearth Kilns

This type of kiln, also referred to as *roller kiln*, is based on the principle of the multi-passage kilns. The bats carrying the ware move on rollers to allow effortless movement, thus avoiding cracking of the bats due to excessive strain. Metal rollers are used for low temperatures, ceramic ones for high temperatures. The idea of using rollers originated in Italy where the first roller kilns were manufactured (Speer, 1981). They are used extensively. They work with both gas and electricity.

4.2.5. Sledge Kilns

In this type of kiln, first brought out in West Germany, the ware moves on special slabs, known as "*sledges*". It is particularly suitable for fast firing hard porcelain "*in-glaze*" decoration to temperatures up to 1280°C (see Chapter 8, section 5, p. 166). Gaseous fuels are used (Harms, 1984).

4.2.6. Walking Beam Kilns

With walking beam kilns, also developed in Germany, there is no need to heat any means of transport. The ware is passed along by ratcheting in a system of alternately moving and stationary beams.

4.2.7. Hover Kilns

This revolutionary type of continuous kiln, developed in Britain, is based on the principle of the hovercraft, the pieces to be fired floating on a cushion of air. Hover kilns distinguish themselves by allowing incredibly short firing times, i.e. fewer minutes than hours in conventional firings. On-glaze decoration has been fired in production hover kilns in 15 minutes (this includes heating and cooling, i.e. total time in kiln). However, it appears that the high cost of suitable refractories for the complicated design has rendered hover kilns too expensive, especially for higher temperatures.

4.3. Comparison of Intermittent and Tunnel Kilns

4.3.1. Assessment of Various Factors

The advantages of general intermittent kilns and conventional high thermal mass tunnel kilns are summarized (by an X) as follows:

Type of kiln	Intermittent	Tunnel
Fuel consumption		X
Maintenance costs		X
Quality of ware		X
Production integration		X
Flexibility	X	
Factory floor space	X	
Capital cost	X	
Social acceptability	X	

Fuel consumption per equal mass of ware is much lower with tunnel kilns than with intermittent ones, because the hot air given off the fired ware in the cooling zone is conducted to the preheating zone to heat the unfired ware. With gas-fired tunnels, the hot air is passed to the burners as secondary air, thus further increasing fuel efficiency. The lower energy requirement with tunnel kilns will be seen from the following data, relating to glost (glazing) firing of earthenware (Jones, 1979):

	kJ/kg
Gas-fired tunnel kiln	5700
Less recuperated gas from cooling zone	630
	5070
Gas-fired *low* thermal mass intermittent kiln	7560

Maintenance costs are much lower with tunnel kilns because the interior walls and roof of the tunnel—being maintained at non-varying temperatures—do not suffer thermal shock, as do the interiors of intermittent kilns as a result of the heating up and cooling down which causes damage to the brickwork. Only the *trucks* going through the tunnel, not the tunnel itself, need any maintenance.

Quality of ware tends to be consistently better in tunnels. This is due to the superior temperature uniformity in tunnel kilns, arising from the much smaller cross-section of tunnel kiln trucks compared with the much larger trucks in intermittent kilns. Differences in temperature are bound to be greater within the larger intermittent kiln trucks. However, it is, of course, possible for superior ware to be obtained from a well-designed and proficiently fired intermittent kiln than from a poorly designed and badly fired tunnel kiln.

Production integration is much easier with a continuous kiln because of its very continuity. Management problems are greatly simplified with a steady flow of ware as guaranteed by tunnel kilns.

Flexibility is one aspect which tunnel kilns lack. They determine the whole factory output (provided there are no intermittent kilns to allow for fluctuations). Both intermittent and tunnel kilns should always be filled to capacity for obvious economic reasons. Leaving gaps in the truck setting, in the case of a short fall of orders, would not only be uneconomical but would also lead to uneven heating and hence imperfectly fired ware. The same thing would occur with slowing down the speed of travel of the trucks through the tunnel. Likewise, a drastic increase in propelling speed in order to increase output would adversely affect the ware quality.

(Filling trucks to capacity and keeping to the correct propelling speed is particularly important for hard porcelain, which requires a reducing atmosphere; otherwise, the smooth gas flow would be upset, intermittent oxidation would take place which would cause blisters and discoloured ware.)

Tunnel kilns thus govern the whole factory organization and cannot be adapted to allow for a boom or recession in trade. (Shutting down a conventional high thermal mass tunnel kiln for short periods would be impractical. It takes weeks, or even months, for it to cool down to room temperature. During holidays the temperature in tunnels is hardly lowered and trucks with "buffer" ware have to be continuously pushed through, although at somewhat reduced speeds.)

The running of intermittent kilns is not affected by fluctuations in trade. If there is a lack of orders, fewer firings are carried out. If there is a boom, full use is made of the total kiln capacity and extra intermittent kilns are built. Being smaller, their erection and installing times are much shorter than those for tunnels.

Factory floor space and its full utilization are important factors in modern industrial management. Several intermittent kilns take up less space than does one tunnel kiln with identical output.

Capital cost involved in the building of intermittent kilns is much less than that of a tunnel kiln designed for an equal amount of ware.

Social acceptability is a highly significant factor. Conventional tunnel kilns have to be run 24 hours a day, 7 days a week and 52 weeks per annum, which means work during the night, the weekend and holidays. Intermittent kiln firing can be done during normal working hours.

4.3.2. *Chronological Development*

Immediately after the demise of the bottle ovens there was a preference for tunnel kilns, mainly because of their lower running costs, less fuel and smaller labour forces having been required. However, the small-scale manufacturers found their high investment cost prohibitive. They chose intermittent kilns which could be built and run in so much more quickly; their number could always be increased if market forces demanded it. Electric intermittent kilns could be fired during periods of cheap, off-peak electricity. Despite their higher running and repair costs, they proved economically and socially more attractive in the long run and the larger pottery factories acquired them in addition to tunnels for the sake of greater flexibility.

A strong case has been made for intermittent firing of bone china by providing concrete evidence that, over a period of 5 years, capital and labour costs for several intermittent kilns were less than for a tunnel kiln giving the same output (Austin, 1976).

Significant improvements have been made, notably in France, in the design of intermittent kilns, leading to savings in energy and, more importantly, to much better temperature uniformity which, in turn, has allowed much faster firing. These advancements have been brought about by the use of fibres for kiln walls and ceiling (see below section 6.2.2), of special burners (see below section 6.2.3) and of *manifold* (electronic) regulation of temperature by means of special dampers; pressure inside the kiln is controlled so that cold air entering the kiln from outside (one of the chief causes of uneven temperature in conventional intermittent kilns) is no longer possible. Moreover, kiln life is prolonged as a result of these measures (Coudamy, 1977).

Yet, despite these advances in intermittent kiln design, there are signs that the pendulum is changing direction again. A new type, the "intermittent tunnel kiln", first conceived as long ago as in the late nineteen-sixties by Holmes (1969) at the British Ceramic Research Association, is likely to prove the ultimate—the last word—in pottery kilns. As the term suggests, it combines the best of both worlds; it will be discussed in section 6.2.1.

5. Conventional Kiln Controls*

The old fireman of the bottle oven era had no instruments to guide him, no means of measuring temperature. As far as bone china and hard porcelain were concerned, he judged the state of the firing by withdrawing samples of ware (so-called "proofs") with a long metal hook from the oven during the final stages of the firing and visually examining them for translucency. When the required standard of translucency had been attained, the firing was stopped. He used his eyes which, due to long experience, were remarkably accurate, for estimating temperatures.

5.1. *Measurement of Temperature and Heat Work*

The potter is not so much interested in actual temperatures as in the combined effect of temperature and time, in other words *heat work*. All heat-work recorders or *thermoscopes* used in the control of firing pottery are based on the shrinkage or deformation of specific ceramic mixtures. For example, the decrease in size of a ring-shaped test piece with increasing heat work is used as a measure of that heat work in *Buller's rings*. The idea was originally developed by Wedgwood (although not adapted until 150 years later). Böttger, the originator of hard porcelain in the Western world, in 1709, was the first to devise recorders of heat work based on deformation. This principle is utilized in *Holdcroft bars* which sag during firing, the more pronounced the sag the greater the heat work.

The most important heat-work recorders, also based on deformation, are *pyrometric cones*, also known by the names of their originators, as *Seger cones* (in Europe) and *Orton cones* (in America). They are slender, three-sided pyramids made of mixtures ranging from low-temperature glazes to pure

---

* The latest sophisticated computer-based controls are dealt with in the next section (Fast Firing) under section 6.2.4, "Microprocessor Controlled Operation".

alumina comprising nearly sixty different compositions, each having an index number corresponding to a certain "melting point" (a measure of the appropriate degree of heat work). This is indicated when the tip of the pyramid has bent down to the same level as the base. The range of "melting" temperature covered extends from 600 to 2000°C.

In modern pottery kiln installations temperatures are measured (and automatically recorded) and controlled by *thermocouples* or *optical radiation pyrometers*. With intermittent kilns a *programme controller* regulates the rate of temperature rise and the duration of the "soak" at the maximum temperature required. Similarly in tunnel kilns temperatures can be maintained constant by automatic adjustments in case of altered composition, calorific value, or pressure of the gas or density in the truck setting.

The thermocouples are fixed at various points in the kiln and do not necessarily give the temperature of the ware on the truck. Exact temperatures of the actual ware can be obtained by attaching thermocouples to the moving kiln cars. However, this is somewhat cumbersome and thermoscopes are therefore still widely used, if only as a check and not as regulators of firing.

5.2. *Measurement of Pressure and Draught*

Correct gas and air pressure and flow, as well as draught, are essential for the conduct of firing. This does not only apply to all types of gas-fired kilns but also to electrically-fired tunnel kilns where hot air has to be conducted away from the exit end of the kiln to the entry end. Ordinary gas flow-meters are used for the measurement of draught.

5.3. *Gas Analysis*

Kiln gases are analysed to ensure efficient combustion and maintenance of the appropriate atmosphere. Pottery is normally fired in an atmosphere of air (normally described as oxidizing). Excess of air is avoided except in the early stages of the firing when carbonaceous matter and combined water from the clays have to be removed.

Knowledge of the composition of the gas is even more important where reducing atmospheres are deliberately introduced as with certain types of stoneware and porcelain.

Kiln gases are analysed with the simple *Orsat* apparatus or more often with one of its developments as an automatic self-recording instrument. Occasionally electrical gas analysis methods find use.

6. Fast Firing

There are no hard and fast rules defining fast firing. The duration of firing depends, of course, on the maximum temperature desired. Some authorities (i.e. Sladek, 1986) regard 2–8 hours, comprising the total time of heating and cooling , as fast firing whereas others stipulate 90 minutes or less (see Figs. 6.6 and 6.7).

FIG. 6.6.   Fast fire tunnel kiln with table conveyance for the glost firing of hard porcelain. By courtesy of Ludwig Riedhammer GmbH.

FIG. 6.7.   Fast fire tunnel kiln with sledge conveyance for the in-glaze and on-glaze firing of tableware. By courtesy of Ludwig Riedhammer GmbH.

6.1. *Theoretical Justification*

It was once thought that slow heating and slow cooling rates were essential to prevent cracking of ware. However, it was established that the cracking which did occur was not due to *rapid* heating and cooling but to *uneven* heating and cooling which caused stresses across the piece. Bone china cups could be biscuit fired in a hover kiln in the incredibly short time of 7 minutes without showing any sign of cracking.

Likewise, plates could be fired singly in a matter of minutes; however, stacks of the same article required hours or even days. The reason why this should be so has been elegantly postulated by Holmes (1969): heat flow (conduction of heat) during firing is the crucial factor; this can be calculated from the *thermal diffusivity* (as opposed to thermal conductivity which would only operate when steady temperatures have been reached).* Diffusivity determines the rate of temperature rise in the centre of an article which is being heated at the surface.

The time required for centres to reach 99.9 per cent of the surface temperature is proportional to the square of the thickness of the piece. Thus, the time required for a slab 2 cm thick is 9.6 minutes, whereas for a slab 20 cm thick it is 96 minutes (1.6 hours); for a plate 6 mm thick it is only 52 seconds. It was shown that the time required for the temperature to be equalized was over 300 times longer for a stack of six tiles than for a single tile.

6.2. *Practical Aspects which Make Fast Firing Possible*

6.2.1. *Kiln Design*

It is obvious from the above facts that the conventional truck-type tunnel kilns would be useless for fast firing. What is required are single-layer kilns. Practically all the types mentioned in the section on tunnel kilns (4.2), except the first one (4.2.1, Conventional High Thermal Mass Tunnels), lend themselves to fast firing. They were, in fact, designed with shortened firing times in mind.

The first intermittent tunnel kiln, built by the pioneers of fast firing (Holmes and others, 1971) could be brought up to temperature in 1 hour. This exceedingly short time was made possible as a result of the very low heat capacity of this kiln, the design features of which have been described in great detail, covering heat transfer, method of conveyance, kiln car superstructure, actual operation, etc. The kiln was designed to operate successfully with several types of fuels. Despite its lightly insulated structure, its thermal efficiency was equal to that of a conventional gas-fired tunnel kiln. Its great

---

* Thermal diffusivity is defined as the thermal conductivity divided by the product of specific heat and density.

asset was that, after having been brought up to temperature in 1 hour, it could be run for an 8-hour shift and then turned off.

6.2.2. Low Thermal Mass Construction Materials

The first conventional tunnel (and truck-type intermittent) kilns had extremely thick walls which were thought to be necessary to prevent heat loss. They consisted of several layers of bricks: highly refractory ones at the kiln face, less refractory ones, insulation bricks and finally high-quality normal facing bricks. Gradually, by introducing high-temperature light-weight insulation bricks, the excessive thermal mass was reduced. The most effective insulation is offered by refractory *fibres*, consisting of alumino-silicates, i.e. mullite, etc. These are now almost exclusively used in intermittent kilns, as well as tunnels. Their chief advantage lies in the fact that they are extremely light (combining low thermal conductivity with low thermal mass) and, therefore, require very little energy in heating up. Their small bulk is an additional bonus. The low heat capacity and excellent thermal shock resistance make fibres an attractive material for fast fire intermittent tunnel kilns.

Fibre materials are not used where there is a danger of chemical attack of the internal kiln lining by glaze condensates; in such cases fibre is replaced by high-quality refractory insulating bricks (Kaether and others, 1983).

6.2.3. Special Burner Designs

Recuperative burners utilize the heat in the waste gases of gas-fired kilns; all the waste gases are taken out of the kiln through the burners. These are combined with metallic counterflow recuperators. The waste gases are drawn out of the kiln through an annulus; some of their heat is transferred to counter-flowing combustion air in an adjacent annulus (Holmes, 1978). The main purpose of recuperative burners is to save energy.

High-velocity burners are designed to create the temperature uniformity across the setting, essential for fast firing. *Jet burners* are the most advanced high-speed burners; they cause the gas to emerge from the burners at such high speeds that it is impossible for hot spots to be formed near the burners. The turbulence caused is responsible for superior heat transfer within the ware being fired (Coudamy, 1977). In other words, the intensive circulation of the kiln atmosphere causes perfect temperature uniformity.

6.2.4. Microprocessor Controlled Operation

The conventional kiln controls, mentioned in the previous section (5), would be of no use for fast firing. To quote Holmes (1981), "already microprocessor based control systems are threatening to consign to science museums the automatic temperature control systems that we once pointed to so proudly and were even beginning to understand". The microprocessor has made it possible to automatically change the firing curves and to programme complete weekend operations. Electronic temperature controllers regulate the flow rate of gas. Oxygen analysers have been developed, based on electrochemical cells, containing a solid zirconia electrolyte. The rapid response of these cells to changes in oxygen content make them ideal sensors in automatic systems for the control of kiln atmospheres. The monitoring of the oxygen content is important in the maintenance of reducing conditions in the firing of hard porcelain and also in order to avoid using excess oxygen in normal oxidizing atmospheres, thus saving fuel.

6.3. *Why Fast Firing?*

The reasons for fast firing can be summarized as follows:

6.3.1. *Cost Savings*

Fast firing requires less *fuel* and hence reduces costs. It has been argued that this *lowering of fuel costs* is counter-balanced by the fact that, because of single-layer firing, several fast fire kilns would have to be installed to produce the same output as one conventional kiln, the capital cost of installing the fast fire kilns being twice as high as for the conventional kiln.

An accurate direct comparison between conventional and rapid firing is almost impossible because of different amortization rates for rapid kilns and conventional ones, and because of other factors, difficult to assess. A special study revealed that fast firing compared favourably with conventional firing for any production capacity (Elias and Poppi, 1980).

Savings in fuel costs are secondary to the benefits to be gained from *single-layer placing* which allows *automation*, the main incentive for adopting fast firing.

6.3.2. *Avoidance of Unsocial Working Hours*

Fast fire intermittent tunnel kilns can be heated within a very short time, i.e. 1 hour or even less, depending on the maximum temperature; this makes it possible for them to be shut off during weekends and at night. Alternatively, weekend programming can be left to the microprocessor.

6.3.3. *Making Decoration Dishwasher-proof*

Coloured decoration, applied on the glaze and fired to the usual low temperature (750–800°C—see Chapter 8, section 5) in the conventional way, suffers bad wear if exposed to treatment in automatic dishwashing machines.

This could be avoided by applying the decoration before glazing, i.e. *under* the glaze. However, only few ceramic colours stand up to the much higher temperatures of the glazing firing. Normal "on-glaze" colours would badly fade at these temperatures.

If, however, the *"on-glaze"* decoration is applied to the glaze in the normal way but then fired *fast* to a high temperature, the decoration "sinks" into the glaze and is, therefore, protected from the dishwasher action; the time at which the "on-glaze" colours are exposed to the high temperature is too short for fading to occur.

This is the one aspect where fast firing has proved unreservedly successful. Far more fast fire kilns are in use for firing hard porcelain *"in-glaze"* ("sink-in") decoration than for any other purpose. Fast firing has made dishwasher-proof most available ceramic colours.

7. Once Firing

The conventional production of pottery involves three firings: biscuit, glost (glazing) and decorating firings. With most types of pottery the first fire (biscuit) is at the highest temperature; no further reactions occur in the body during the subsequent—lower temperature—glazing fire; there is no more shrinkage and hence no supports are needed for the glost fire. With hard porcelain and certain studio pottery the biscuit fire is at a low temperature (less than 1000°C); it is just high enough to give the pieces sufficient strength for withstanding dipping in the glaze suspension; the glost fire is at the highest temperature, all reactions taking place then, involving softening of the body and shrinkage; therefore, the necessary precautions, cited under section 2.1, have to be taken. With all pottery the traditional decorating fire is at a low temperature (750–800°C). (The decorating fire, also referred to as the "enamel" fire, is sometimes omitted, for instance with under-glaze or in-glaze decoration or when coloured glazes and so-called "slip" decoration provide the desired embellishment.)

The question arises: Are three fires really necessary? Why not just one? There are two weighty reasons for having three fires.

(a) For many years dipping the ware in an aqueous suspension of the finely divided glaze materials has been the quickest and in all ways the most satisfactory method of glazing. Dipping unfired pottery pieces in a glaze slip would cause them to disintegrate immediately unless the pieces were excessively thick and made up of a high percentage of a clay with an exceptionally high green strength.

(b) Prior to the introduction of in-glaze decoration, once-firing would have put severe limitations on the range of colour decoration, as only very few ceramic colours stand up to high temperatures.

These two reasons are no longer fully valid, as more advanced methods of glazing, viz. spraying, have been introduced and as in-glaze decoration has extended the high temperature range of ceramic colours.

The advantages of once-firing are as follows:

1. Cost reduction through fuel savings by eliminating the two lower temperature firings.
2. Omission of capital expenditure of the two lower temperature kilns.
3. No need to keep semi-finished stock (saving of floor space).
4. Lower labour costs, as less handling of ware and hence:
5. Fewer breakages and other damage to ware.
6. General savings arising from simpler flow line and more streamlined production.
7. Dishwasher-proof decoration.

The disadvantages are:

1. Design constraint imposed by the ware having to be thick to withstand the action of spraying (no matter whether done by hand or automatically).
2. The need for the ware to have to be supported during firing as in the glost fire of porcelain.
3. Likelihood of glaze faults through reaction between glaze constituents and enormous volumes of water, liberated in the course of the kaolinite lattice break-up.
4. Limitation of decoration to high-temperature colours, unless fast firing is applied beyond the critical temperature of the break-up of the kolinite lattice.

The firing of "clay-glazed" cups has been described by Clough and Simcock (1976). Once-firing has been introduced in plants manufacturing the cheap mass production type of pottery. However, one large factory in Finland produces high quality once-fired stoneware. This type of pottery is particularly suitable for once-firing because the body has a high green strength (which even allows *dipping* in glaze of unfired hollow ware, difficult to spray) and suitable suction properties (Remmer, 1980).

Further Reading

General: Singer and Singer (1963, pp. 167–227, 854–1047); Holmes (1978).
The effect of heat on ceramics: **Ford (1967)**.
Historical: Rhodes (1971).
Heat transfer, thermal efficiency: Dinsdale (1986, pp. 157, 159).
Sintering theory: Coble and Burke (1963, pp. 199–251); White (1965).
Sintering in solid-liquid system: Kingery (1959, 1960, 1961).
Clay-feldspar-quartz system: **Lundin (1959)**; Schüller (1964, 1979).
Calcium phosphate-containing system. **Beech (1959)**; **Dinsdale (1967)**.
Phase diagrams: Levin and others (1964).
Kiln furniture: White (1953); Lovatt (1984); Harms (1976, 1979).
Fuels: Dinsdale (1953); Metzel (1962); Holmes (1981).
Kilns: Dinsdale (1953); Coudamy (1977); Kaether and others (1983); Harms (1984).
Kiln operating costs: Elias and Poppi (1980); Austin (1976); Holmes (1981).
Thermal technology for tunnel kiln firing: Gardeik and Scholz (1981).
Fast firing: **Homes (1969, 1981, 1984)**; Holmes and others (1971); Speer (1981); Sladek (1986).
Once-firing: Remmer (1980).

7

Glazing

1. Definition and Origin

Glazing is the process of covering the body with a thin layer of glass. A suspension (slip) of the finely ground constituent glaze materials is applied to the body which is then dried and fired. The glassy state is developed during firing. The purpose of glazing is to make pottery more pleasing to the touch and eye and to provide an impervious coating (in case of a porous body) which makes the ware more hygienic, more resistant to chemicals, and in many cases mechanically stronger.

The earliest glaze known is that in stone beads of the Badarian age in Egypt, about 12,000 B.C. Accordingly, glaze is 5000 years older than pure glass, although the glaze referred to was probably very different from what is now used.

2. The Structure of Glasses and Glazes

It is only fairly recently that the pottery technologist has considered glazes from a fundamental point of view. To simplify matters he disregards the fact that the glaze is present only as a thin layer and fused to (and thus united with) the body. He considers the glaze as what it really is, namely a glass, even though a special glass, falling within a fairly restricted range of the wide field of glass structures.

In his classical work on glass structure, Zachariasen (1932) found that: "the atoms in glass must form an extended three-dimensional network . . . which . . . is not periodic and symmetrical as in crystals." Broadly speaking, crystals represent an ordered system whereas glasses are disordered, although the glass network is not entirely random (see Figs. 7.1 and 2.16).

There are a number of compounds which can form glasses. These are termed "network formers". Those of practical importance for pottery glazes are silica and boric oxide. With both oxides the structural units (SiO_4^{4-} and BO_3^{3-} respectively) are joined together at their corners by sharing oxygen ions to form an irregular three-dimensional network. Fused or vitreous silica and fused boric oxide are, in fact, the simplest glasses. Fused silica consists of tetrahedra and is strong, whereas fused boric oxide, being made up of only triangles, is weak.

124

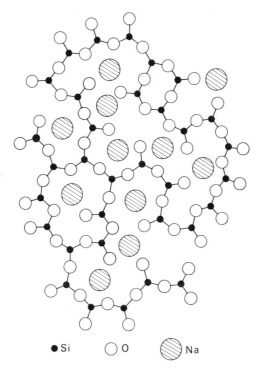

● Si   ○ O   ⊘ Na

FIG. 7.1. The structure of soda-silica glass. (After Hauth (1951), p. 204.)
Reproduced by permission of The American Ceramic Society.

However, neither of these fused oxides would be any use as glazes because
fused silica has too high a melting point (above 1700°C) and fused boric oxide
by itself is soluble in water. Nevertheless, silica is the basis of all pottery
glazes and boric oxide is present in the majority of them. To cause them to
melt at a reasonably low temperature their structure is modified by so-called
"network modifiers". Amongst these are the larger cations, e.g. alkalis and
alkaline earths. Certain cations, such as sodium, cause gaps in the network of
the glass (see Fig. 7.1). This weakens the glass structure and causes
devitrification or poor chemical resistance. However, if another cation, say
CA^{++}, is introduced in addition to Na^+, the gap or gaps in the structure are
closed and a much stronger structure is produced.

In borosilicate glasses consisting of one molecular part of boron to four
molecular parts of silicon, boron becomes four co-ordinated and hence
contributes to strength.

Alumina in the presence of alkali and/or alkaline earth oxides can take over
the function of boric oxide, again forming tetrahedra, co-ordinated by oxygen
ions taken from the alkalis, lime, etc. In certain cases alumina helps to close
the gaps if it replaces silica. Alumina and a number of other oxides, namely

those of Pb^{2+}, Ti^{4+}, Zr^4, and sometimes also the alkaline earths Mg^{2+} and Be^{2+} which behave similarly, are called "network co-formers".

Zinc oxide forms its own tetrahedra, each closely associated with two alkali ions, and causes these to be held more strongly than they would be in ordinary interstices surrounded entirely by SiO_4 groups. Therefore chemical resistance is increased by zinc oxide.

Lead is joined by two oxygens which are corners of two different SiO_4 tetrahedra. High amounts of lead oxide can be introduced in lead borates and especially lead silicates. The Pb^{2+} ions are capable of forming "bridges" between the SiO_4 groups to a far greater extent than other divalent ions.

3. Body-Glaze Relationship

3.1. *The Significance of Thermal Expansion*

The property which matters most in "matching" body and glaze is *thermal expansion*. One would expect that, to obtain the optimum "fit", body and glaze should have identical expansions. However, this is not so. With practically all types of pottery the glaze expands less than the body. There is a good reason for this. Ceramic bodies and glazes, unlike metals, are much stronger in compression than in tension. Therefore the weaker link of the two, namely the glaze, is put under compression. To bring this about a glaze is chosen with an expansion lower than that of the body. During cooling the glaze shrinks less than the body and therefore the glaze is compressed.

If the expansion of the glaze were higher than that of the body, the glaze would shrink more than the body during cooling and would, therefore, be put in tension. This leads to what is known as "crazing" (fine hair cracks in the glaze similar to "crazy paving").

It is relatively easy to alter the thermal expansion of the body by changes in composition (silica and alkali contents), fineness of the raw materials, and firing temperature. On the other hand, the thermal expansion of glazes for a particular temperature range is defined within narrow limits and cannot be changed materially by composition or any other means. Therefore, if mismatch exists between body and glaze, it is always expedient to change the body rather than the glaze.

There are special bodies for cooking ware with very high thermal shock resistance, such as lithia and cordierite bodies (see Chapter 9, section 1.4, p. 175 and section 2.3, p. 193). This special characteristic is due to their exceedingly low thermal expansions. Such bodies are very difficult to glaze because there are no glazes in the accepted sense which have sufficiently low expansions.

However, compositions in the system lithia (Li_2O)—zinc oxide (ZnO)— alumina (Al_2O_3)—silica (SiO_2), applied to beta spodumene ($Li_2O.Al_2O_3.$ $4SiO_2$) bodies having coefficients of thermal expansion of 2×10^{-6} (20–

1000°C), gave satisfactory matt glazes without crazing or any other fault. This was brought about by subjecting the pieces to a special heat treatment, not unlike that employed in the manufacture of glass ceramics (see Chapter 9, section 2.4, p. 194), viz. maturing at 1250°C ± 30K for 10–30 minutes, cooling to 550–650°C at a rate higher than 22K/minute and reheating at 750–1000°C for 1–6 hours. The glaze had good chemical durability and the articles coated with it showed good thermal shock resistance (Takeda and others, 1974).

3.2. The Influence of Glaze on the Development of Pottery Bodies

It is fair to say that the glaze largely governs the composition of the body. This is illustrated by two exmples.

The first concerns earthenware. In Chapter 2 (p. 27) reference was made to the deleterious influence of crystalline silica. Its replacement by other fillers in earthenware (all of which have much lower expansions than crystalline silica materials) causes a marked improvement of several properties. However, the resultant body no longer fits the glaze because its expansion is lower than that of the glaze; hence the glaze crazes. Crystalline silica is therefore still used in earthenware.

The second instance where the glaze dictates body composition is shown with porcelain. Looking at Table 1.1 (Chapter 1, p. 3) the reader may have wondered why two types of pottery similar in composition, vitrified hotelware and hard porcelain, have such widely different firing treatments. In view of the temperatures of the glazing fire, vitreous china requires a low-temperature glaze, hard porclein a high-temperature glaze. Low-temperature glazes have high expansions, high-temperature glazes have low expansions. The vitreous china body (like earthenware) has a high expansion by virtue of its high quartz content. Therefore it requires a low-temperature (high expansion) glaze. If it had a high-temperature glaze (with a corresponding high-temperature glazing fire) the compression of the glaze (caused by its low expansion as against the high expansion of the body) would be exceedingly high; the body would be greatly weakened as it would be under excessive tension. As a result the glaze would disrupt the body. This feature is called "chittering" and in extreme cases the ware disintegrates, splitting up into small bits, either during cooling in the kiln or some time after withdrawing it from the kiln.

Low-temperature glazes are much more brilliant than high-temperature ones. If (for this reason) hard porcelain were first fired to its final high temperature (1400°C), then glazed with a low-temperature glaze (followed by the glazing firing at low temperature) the glaze would craze. This is so, of course, on account of the expansion of the glaze being higher than that of the body.

TABLE 7.1. *Effect of Variations in the Firing Temperature on Body/Glaze Fit of a Vitreous China Body*

	1	2	3	4	5[a]
Type of glaze	Low-temperature glaze	Low-temperature glaze	High-temperature glaze	High-temperature glaze	Low-temperature glaze
Firing temperature: First fire (biscuit)	<1380°C	>1380°C	900°C	900°C	1280°C
Glazing fire	1100°C	1100°C	>1400°C	<1400°C	1100°C
Coefficient of thermal expansion ($\alpha \times 10^{-6}$) Body	>7.0	<7.0	<5.5	>5.5	7.4
Glaze	7.0	7.0	4.4	4.4	7.0
Body glaze fit	Perfect	Crazing	Perfect	Chittering	Perfect

[a] Normal treatment of vitrified hotelware body.

Hard porcelain distinguishes itself by its bluish-white hue (due to the reducing fire), vitreous china by the brilliance of its glaze. In view of thermal mismatch it seems impossible to combine these attractive features, but by very careful control of the firing temperature it can be done. The effects of variation in the firing temperature fare shown in Table 7.1.

Bodies within the vitreous china composition can be fired first (reducingly) to a high temperature which brings out the characteristic bluish tint normally associated with hard porcelain. However, the temperature must be such that sufficient crystalline silica is left in the body to ensure that it has an expansion high enough to match that of the low temperature glaze. As little as 20–30°C is sufficient to dissolve this critical amount of silica, thus eliminating the quartz kink in the expansion curve of the body and lowering its overall expansion. Thus a body "biscuited" (which means fired unglazed) at 1380°C may be perfectly suited to carry a low-temperature brilliant glaze without crazing (see 1 in Table 7.1), whereas the same body biscuited at 1400°C causes crazing of the same low-temperature glaze (see 2 in Table 7.1). On the other hand (having been biscuited at 900°C in the manner of hard porcelain) this body is suitable for taking a normal low expansion–high temperature (less brilliant) hard porcelain glaze if fired to 1400°C or above (see 3 in Table 7.1). If the same body were again glazed with a hard porcelain glaze but fired to 1380°C or less the result would, of course, be chittering (see 4 in Table 7.1).

3.3. *Interaction of Glaze and Body*

As the glazing firing approaches its peak temperature the glaze becomes fluid and attacks the body. Interaction layers are formed in the glaze–body interface. These may consist of crystals which have expansions different from either body or glaze, e.g. large, needle-shaped crystals of (secondary) mullite in the normal triaxial system or apatite in bone china, the phosphorus pentoxide (P_2O_5) reacting with lime (CaO) in the glaze, also affecting the refractive index. These interlayers further influence and complicate body–glaze relationships. Glaze–body reactions on bone china and earthenware were studied by means of electrochemical measurements over a range of temperatures (Roberts and Marshall, 1970). Reaction was found to be due to the movement of O^{2-} ions, accompanied by that of mobile cations. The major contribution to the reactions came from mullite, as far as earthenware was concerned, and from the body-glass in respect of bone china. These results were verified by electron-probe micro-analysis.

In a further body–glaze reaction study on earthenware it was tried to ascertain to what degree the interfacial layer between body and glaze influenced the formation of craze marks. The effect of various oxides in the glaze was investigated at a constant glaze viscosity. Larger amounts of lead oxide (PbO), zinc oxide (ZnO), magnesia (MgO) and lithia (Li_2O) caused more active body–glaze reaction, forming pronounced interlayers and increasing the tendency to crazing: baria (BaO), potash (K_2O), strontia (SrO) and boric oxide (B_2O_3) proved harmless: lime (CaO) and soda (Na_2O) occupied intermediate positions.

4. Glaze Compositons

The knowledge of the fundamental mechanism of glass formation has helped the potter a great deal in explaining certain phenomena in glazes, but he would be unwise to rely solely on it in devising new glazes.

4.1. *The Concept of Molecular Formulae*

In the past, at least up to 100 years ago, the compositions of glazes were arrived at by trial and error, if not accident. In bodies the range of composition was fairly well defined but with glazes there existed an enormous number of odd mixtures apparently bearing no relation to each other. It was the great merit of Hermann Seger to have brought order into the chaos of glaze compositions. He was perhaps the first (after the valiant attempts of Josiah Wedgwood 100 years before him and perhaps Bernard Palissy 200 years before Wedgwood) to succeed in bringing science into the pottery industry. He conceived the idea of representing glazes as chemical formulae. It is interesting to note that the former mixtures of more or less mysterious ingredients follow certain well-defined laws and rules which must have been unknown to their originators.

The shape of the formula is quite simple, the oxides being listed in three columns, viz.

$$RO \quad R_2O_3 \quad RO_2,$$

where RO represents the sum of basic oxides (taken arbitrarily as unity), R_2O_3 the neutral oxides, and RO_2 the acidic oxides. Expressed differently, the first column contains the monovalent and divalent elements, the second the trivalent, and the third the tetravalent elements.

The basic or monovalent and divalent elements include the alkalis and alkaline earths, the acidic or tetravalent elements comprise silica (silicic acid) only—although, based on the "acidic" concept, boric oxide (boric acid) was grouped by Seger with silica, but has been moved by Singer (1910) to the trivalent column, appearing together with alumina.

Taking Seger's original concept, all the network modifiers are in the first column, the network formers in the third column, with the network co-formers in the middle. Seger's arrangement was thus in a way prophetic because in his day nothing was known of the fundamentals of glass structure.

Seger's molecular formulae are not the only means of presenting glaze compositions. Some ceramists prefer the weight percentage composition or the molecular percentage composition of the various oxides. The majority of pottery technologists regard Seger's presentation still as the most useful, logical, and easiest to survey, especially when comparing glazes.

4.2. *Fluxing Action of Oxides Making Up Glazes*

The bases (network modifiers) invariably act as fluxes. Of the acidic oxides (network formers), boric oxide also acts as a flux. Silica increases the melting temperature and so does the "neutral" (network co-former) alumina.

Regarding the network modifiers the fluxing effect of the monovalent oxides (alkalis) is the greater the smaller the cation. Thus soda is more effective in reducing the melting temperature than potash, and in turn lithia is more effective than soda.

Taking the bases as a whole, their fluxing effect is as follows in decreasing order:

$$LiO_2 - PbO - Na_2O - K_2O - BaO - CaO - SrO - MgO - ZnO.$$

Replacement of silica by boric oxide decreases the melting temperature, as does a decrease in the silica and alumina contents. A further lowering of the fluxing temperature is achieved by complexity, viz. by the introduction of a greater number of bases. Thus, in practice, glazes are far more complex than glasses, which usually only contain three or four oxides.

Glazes are further complicated by being present as thin layers (low mass), their reaction with the body, and their non-glassy inclusions. For these reasons phase diagrams (as in the case of bodies) are of limited use, although reactions are more or less taken to completion in glazes.

4.3. *Typical Glazes*

Let us start at the high-temperature end (1400°C) and consider what might be regarded as the simplest glaze, simple because it is made up of one single mineral, potash feldspar.

Glaze 1. $K_2O.Al_2O_3.6SiO_2$ Maturing temperature* 1400°C

Such a glaze (which, incidentally, could only be applied to hard porcelain since no other type of pottery is fired to such a high temperature) would craze because of thermal mismatch with the body. It would also lack brilliancy. These faults are due to insufficient silica as well as the fact that only alkali was used as a base.

Lime as the only base produces a suitable glaze for temperatures above 1380°C, but normally RO is made up of two or more *basic* oxides. The ratio

---

* *Maturing temperature* is the temperature to which the glaze has to be fired to ensure that it has reached optimum brilliancy.

alumina/silica usually varies between 1 and 8–10. If in the above glaze (1) a large portion of potash is replaced by lime and if silica is increased the result is a sound, hard porcelain glaze, viz.

Glaze 2. $\left. \begin{array}{l} 0.7\,CaO \\ 0.3\,K_2O \end{array} \right\}$ $Al_2O_3.10\,SiO_2$ $\qquad\qquad$ 1400–1440°C.

Incidentally this is the formula for Seger cone 10 (see Chapter 6, p. 115). The composition of Seger cones within a certain range can indeed be used as glazes for porcelain provided they are fired four to five cones (about 100°C) higher than the melting temperature of the cone concerned. Usually the alumina/silica ratio in glazes is not as high as 1:10 (as in the formulae for Seger cones) but only 1:9 or less.

Descending to the porcelains fired at lower temperatures, the so-called soft porcelains or hard stoneware, the following formula gives a good glaze:

Glaze 3. $\left. \begin{array}{l} 0.2\,ZnO \\ 0.2\,MgO \\ 0.4\,CaO \\ 0.1\,K_2O \\ 0.1\,Na_2O \end{array} \right\}$ $0.4\,Al_2O_3.3.5\,SiO_2$ $\qquad$ 1220–1260°C.

Not only are alumina and silica reduced, but the number of bases is increased (although there is no reason why the higher-temperature glaze should not have five instead of two bases).

In glazes—required to mature below 1200°C—for bone china, vitreous china, stoneware, earthenware, majolica, or common pottery, lead oxide (PbO) and boric oxide (B_2O_3) are rarely absent. Lead oxide adds great brilliance to the glaze. Whereas glazes without lead oxide have been devised, boric oxide is now invariably used in glazes maturing below 1200°C.

Examples of leadless glazes are:

Glaze 4. $\left. \begin{array}{l} 0.2\,MgO \\ 0.2\,CaO \\ 0.3\,BaO \\ 0.3\,K_2O \end{array} \right\}$ $\left. \begin{array}{l} 0.4\,Al_2O_3 \\ 0.5\,B_2O_3 \end{array} \right\}$ $3.5\,SiO_2$ $\qquad$ 1150–1170°C.

Glaze 5. $\left. \begin{array}{l} 0.1\,MgO \\ 0.7\,CaO \\ 0.1\,K_2O \\ 0.1\,Li_2O \end{array} \right\}$ $\left. \begin{array}{l} 0.3\,Al_2O_3 \\ 0.7\,B_2O_3 \end{array} \right\}$ $3.5\,SiO_2$ $\qquad$ 1140–1160°C.

The high fluxing power of boric oxide is apparent when comparing glazes 3 and 4. Both have the same alumina and silica values, yet the one with boric oxide (glaze 4) matures about 80°C lower; the baria content of glaze 4 also helps to reduce the maturing temperature.

Proceeding to the lead glazes we can keep to the same alumina and silica contents as in glaze 5, and even a lower boric oxide content than in glaze 4, yet produce a glaze maturing at a lower temperature by the introduction of lead oxide as in:

Glaze 6. 0.3 PbO
0.1 MgO
0.4 CaO   } 0.3 Al_2O } 3.0 SiO_2 1080–1120°C.
0.1 K_2O } 0.4 B_2O_3
0.1 Na_2O

To further reduce the maturing temperature alumina and silica are lowered and lead oxide is increased, e.g.

Glaze 7. 0.7 PbO
0.1 MgO   } 0.2 Al_2O_3 } 2.0 SiO_2 950–970°C
0.2 CaO   } 0.4 B_2O_3

As we approach the lowest maturing temperatures, glazes become simpler again and only one base, the most powerful flux (apart from lithia), viz. lead oxide, is used, which allows boric oxide to be eliminated (there are glazes for the 1000–1200°C range which are also free from boric oxide).

Glaze 8. $PbO.0.1 Al_2O_3.1.0 SiO_2$ 900°C.

Pottery glazes are rarely applied at temperatures below 900°C. However, glazes for such low temperatures exist, the simplest being lead bisilicate:

Glaze 9. $PbO.2 SiO_2$ 700–800°C.

Compositions of glazes maturing at 650°C and even 550°C have been published (Singer, 1954). Glazes based on P_2O_5 as the only network former melt at 400°C but are soluble in water.

5. Special Glazes*

5.1. *Glazes with a Crystal Phase*

Normally, glazes are clear, transparent glasses of high brilliancy. Sometimes special effects are produced which, like so many aspects of pottery,

* Coloured glazes are discussed in Chapter 8, "Decoration", (section 4.3, p. 158).

were probably first discovered by accident. These types of special glazes contain a second phase.

5.1.1. *Opaque Glazes*

Opacity in glasses or glazes is obtained by introducing fine crystals insoluble in the glaze, and of different refractive index. It can also be brought about by crystallization of the melt or by introducing substances which are immiscible and thus form two vitreous phases, as with a mixture of phosphate and silicate glasses which separate on cooling.

The opacity is the greater the more often the path of light is broken by the foreign crystals; the finer the crystals the greater the surface area and hence the greater the reflection. This is only correct up to a certain fineness of particles. Particles of the same order as the wavelength of light are unable to arrest the light rays. The particle size for optimum opacity is about 0.4 μ.

Tin oxide (SnO_2) used to be most widely employed opacifier but has now been largely replaced by Zircon ($ZrO_2.SiO_2$) which is much cheaper. Zirconium compounds are inert, whereas other opacifiers, including tin oxide, react with some of the glaze constituents. This applies to titania (TiO_2), which enhances opacity if other opacifiers are present.

With the higher temperature glazes, as used for hard porcelain, compositions rich in alkali and alumina (the alumina/silica ratio rising to 1:7) appear opaque. Porcelain glazes low in alkaline earths, in any case, tend to be more opaque than the more brilliant lower temperature lead-borosilicates.

5.1.2. *Matt Glazes*

Opaque glazes can be regarded as a half-way stage to matt glazes. The matt effect is again due to small crystals. It is the result of devitrification produced when a completely fused glaze cools and part of the fused mass crystallizes. The crystals are very small and evenly dispersed, giving the glaze surface a smooth and velvety appearance. Matt glazes are always opaque because the crystals, as in normal opaque glazes, break up rays of light.

The formation of the small crystals producing the matt surface can only occur if sufficient time is available. Matt glaze compositions cooled too rapidly have glossy surfaces.

The crystallization on cooling is usually obtained by increasing the alumina content. Silica is reduced in favour of alumina, the alumina/silica ratio being within 1:6–1:3. Sometimes barium oxide produces matt surfaces, the crystals being present as celsian (barium feldspar $BaO.Al_2O_3.2SiO_2$). In matt glazes containing lime the crystals are anorthite (lime feldspar $CaO.Al_2O_3.2SiO_2$).

5.1.3. *Crystalline Glazes*

In this type of glaze the crystals, contrary to opaque and matt glazes, are very large so that they can be seen with the naked eye. Very beautiful effects can be produced by supersaturating the glaze with a compound that crystallizes easily when cooled very slowly.

The constituents capable of producing crystalline glazes are zinc silicate, zinc titanate, manganese silicate, and calcium and magnesium silicates. Tungsten trioxide (WO_3), vanadium pentoxide (V_2O_5), and molybdenum trioxide (MoO_3) also produce crystals in glazes.

The mechanism operative in giving crystalline glazes is similar to that which produced Réaumur's porcelain and glass ceramics (see Chapter 9, p. 193). Nuclei are formed during cooling which lead to crystal growth usually at a temperature somewhat higher than the actual nucleation temperature. Glass ceramics and crystalline glazes differ only in the size of their respective crystals; opaque and matt glazes with their minute crystals are, in a way, nearer to glass ceramics than crystalline glazes.

5.1.4. *Aventurine Glazes*

These are a special type of crystalline glaze. Their crystals are plate-like, contrary to the needle-shaped crystals of normal crystalline glazes. They are also smaller and more uniform than those in ordinary crystalline glazes. Aventurine glazes are nearly always associated with colour effects and they usually contain the oxides of iron, chromium, or copper.

Crystal growth again depends on careful cooling and to some extent on the manner of heating.

5.2. *Glazes with Special Effects Caused by the Body*

5.2.1. *Craquelé (Crackle) Glazes*

In this type of glaze a fault is made into a virtue. Crackle glazes are the result of thermal mismatch between body and glaze. As mentioned, a glaze crazes (cracks) when its thermal expansion is higher than that of the body. We have already come across the simplest example: feldspar as a glaze on hard porcelain (see p. 131 of this chapter).

The effect of the craze pattern is often enhanced by rubbing colouring matter, such as amber, into the cracks.

5.2.2. *Snakeskin Glazes*

As mentioned previously, the thermal expansion of a normal glaze is lower than that of the body. If, however, the difference between the two expansion rates is excessive the glaze shows "peeling" or "crawling". This can have the appearance of the skin of a snake.

Similar effects can be produced by high surface tension of the glaze. This is achieved by inducing high shrinkage, viz. applying the glaze very thickly, adding a large amount of clay to it, and overgrinding.

5.2.3. Salt Glazes

Salt glazes are applied to a certain type of stoneware. They are not produced in the normal manner of applying a finely dispersed suspension of glaze materials to the body in the cold state followed by firing. The salt glaze is formed on the biscuit body by reaction of common salt with the body constituents, particularly silica, towards the end of the firing.

The body must be richer in silica than normal stoneware. Iron impurities in the body help to give good salt glazes. They tend to vitrify the body, which, in turn, assists glaze formation. A reducing firing is often employed as the reduced iron silicates are very powerful fluxes. The iron content of the body also governs the temperature of salting. It is not usually below 1150°C even with bodies rich in iron, and may rise to above 1300°C with bodies consisting of normal stoneware clays.

Salt glazes are produced in periodic kilns fired by coal, gas, or oil. The salting mixture, consisting of sodium chloride and water, is introduced into the kiln when the appropriate temperature has been reached. The damper is closed for a short time and then reopened. When the vapours have cleared the next salting takes place. Four to eight saltings may be made. Cooling is usually in oxidizing conditions.

During salting the sodium chloride is vaporized and in reacting with steam forms hydrogen chloride and soda. This reacts with the silica in the body and other body constituents. A glaze having a fairly high alumina content (0.6 molecular parts) and a relatively low silica content (2.6 molecular parts), and in which the main base is soda, is formed.

Salt glazes have been improved by the addition of borax and sometimes sodium nitrate to the salting mixture. Often colouring oxides are incorporated in the salting mixture giving beautiful effects, occasionally a kind of aventurine glaze.

5.2.4. Self-glazing Bodies

Certain types of bodies in the magnesia–alumina–silica system (e.g. cordierite-type bodies) are self-glazing at their normal firing temperatures (or at higher temperatures) without the influence of vapours as with salt glazing. They never show high gloss and are mostly matt.

Parian bodies (see Chapter 9, p. 184) also produce a certain glossiness.

6. Raw Materials for Compounding Glazes

For reasons of economy glazes are not made up by mixing the separate oxides mentioned under 4. "Glaze Compositions" but, as far as possible, from naturally occurring minerals. They are largely the same materials as introduced into bodies and conveniently are often the source of more than one constituent oxide. Table 7.2 contains the most common materials used. Soluble materials are marked **S**.

7. The Processing of the Glaze

Glazes for high temperatures, say above 1200°C, as used in hard porcelain, are prepared simply like mill mixtures of bodies. Lower temperature glazes require "fritting" which means premelting followed by cooling.

7.1. *Fritting*

Certain ingredients, especially boron compounds, as needed for low-temperature glazes, are soluble in water. This means that the normal methods of preparation (water grinding) and application of glaze as slip (aqueous suspension) would be impossible. For this reason the water-soluble ingredients are made insoluble; they are mixed dry with all or some of the other glaze materials and melted to a glass-like substance, the "frit".

There are two further reasons for fritting. It makes poisonous lead compounds non-toxic and therefore performs an important function in protecting the health of the glaze workers.* Furthermore, with fritting (premelting the materials) the reactions having already been taken to completion, less heat work is required in the eventual firing of the glazed ware. This means that the surface of a fritted glaze is superior to that of a raw glaze of identical chemical composition fired at the same temperature.

The industrial process of fritting is continuous. The finely grained materials are mixed dry and fed to a simple glass-melting furnace at a controlled rate. The melt runs into open tanks filled with water causing it to break up into globules of glass about 2–25 mm (¹⁄₁₆–1 in.) in size. In modern plants, water quenching is replaced by an arrangement producing very thin platelets. This method gives a more uniform product and is cheaper as it eliminates the need for drying. Besides, the thin flakes are more easily ground than the water-quenched globules.

To obtain optimum glaze qualities all materials are fritted together, excepting a portion of china clay, not exceeding 10 per cent, which helps to keep the glaze in suspension when ground.

---

* According to British factory regulations the amount of soluble lead in glazes must not exceed 5%. Such glazes are called "low sol" glazes.

TABLE 7.2. *Glaze Raw Materials*

Oxide desired	Raw material	Other oxides introduced
Silica (SiO_2)	Crystal quartz, quartz sand, flint	—
	China clay	Al_2O_3
	All feldspathic and micaceous minerals	K_2O, Na_2O, Li_2O, and Al_2O_3
	Wollastonite	CaO
	Talc	MgO
	Zircon	ZrO_2
Alumina (Al_2O_3)	China clay	SiO_2
	All fedspathic and micaceous minerals	K_2O, Na_2O, Li_2O, and Al_2O_3
	Corundum	—
	Aluminium hydrate	—
Potash (K_2O)	Potash feldspar	Al_2O_3 and SiO_2
	Nepheline syenite, Cornish stone	Na_2O, Al_2O_3, and SiO_2
	Potassium nitrate S	—
	Potassium carbonate S	—
Soda (Na_2O)	Soda feldspar	Al_2O_3 and SiO_2
	Nepheline syenite, Cornish stone	K_2O, Al_2O_3, and SiO_2
	Sodium carbonate S	—
	Borax S	B_2O_3
Lithia (Li_2O)	Spodumene, petalite	Al_2O_3 and SiO_2
	Lepidolite	K_2O, Al_2O_3, and SiO_2
	Amblygonite	Al_2O_3, P_2O_5
	Lithium carbonate	—
Lime (CaO)	Wollastonite (rarely)	SiO_2
	Lime feldspar (very rarely)	Al_2O_3, and SiO_2
	Calcium carbonate, viz. chalk, limestone, marble	—
	Dolomite	MgO
	Calcium borate	B_2O_3
Magnesia (MgO)	Talc	SiO_2
	Magnesium carbonate, viz. magnesite	—
	Dolomite	CaO
Baria (BaO)	Barium feldspar (Celsian, very rarely	Al_2O_3 and SiO_2
	Barium carbonate	—
	Barium sulphate (barytes)	—
Strontia (SrO)	Strontium carbonate	—
Zinc oxide (ZnO)	Zinc oxide	—
	Zinc carbonate	—
Lead oxide (PbO)	Lead oxide, red lead, lead dioxide, lead carbonate etc.	—
Boric oxide (B_2O_3)	Boric oxide S	—
	Borax S	Na_2O
	Calcium borate	CaO
Tin oxide (SnO_2)	Tin oxide	—
Zirconia (ZrO_2)	Zirconia	—
	Zircon	SiO_2

7.2. *Grinding*

With glazes requiring fritting the water-soluble materials are invariably supplied in powder form. If the non-soluble materials, e.g. feldspar and quartz, are not supplied in a fine form they have to be crushed, ground, and magneted. Drying, pulverizing, and sieving prior to mixing is necessary with water-ground materials.

The prepared frit, china clay, and possibly other "mill additions" (materials not incorporated in the frit) are ground in ball mills with water, followed by sieving and magneting as described for the non-plastic body materials (see Chapter 3, p. 41). Non-fritted glazes are treated in the same way.

Sometimes ground glaze slips have a tendency to form hard sediments. This applies particularly to non-fritted glazes high in feldspar, a minute portion of which may become soluble and act as a strong deflocculant. Flocculants, such as calcium chloride, not exceeding 0.1 per cent of the mill charge, serve to prevent this sedimentation. Occasinally bentonite and organic compounds are used as suspending agents.

Glazes are ground more finely than bodies, i.e. to 80 per cent less than 10 microns; the greater the fineness of the glaze batch, the better the surface of the fired glaze. Grinding in vibro-energy mills (see Chapter 3, section 2.2.3, p. 42) has been recommended, resulting in a major saving in processing time, viz. 2½ hours compared with 18 hours' ball milling (Slinn and Rodgers, 1980); the narrow range of particle distribution is said to contribute to the improved surface. Excessive fineness causes the glaze to dry slowly after it has been applied to the biscuit pieces, and to "crawl" (see section 5.2.2., "Snake-skin Glazes", p. 135) during firing.

After grinding, the glaze slip is sieved and magnetted.

8. Conditioning of Ware for Glazing

Prior to applying the glaze the fired ware is in what is termed the "biscuit" state. Porous ware is "brushed" in order to remove traces of dust. With ware of low or no apparent porosity the placing medium in the biscuit firing, e.g. alumina, may stick to the ware, and if so, the pieces have to be "scoured" or rumbled. Scouring is brushing with mechanized, very stiff brushes. Rumbling takes place in large, wooden boxes furnished with shelves revolving horizontally round their own axes. The articles are placed in these shelves and while the boxes revolve broken bits of biscuit ware ("pitchers") rub against the articles, which thereby obtain a smooth surface.

9. Application of Glaze

The prepared glaze slip is normally applied to the *fired* pottery body. Thin-walled, delicate pieces would be mechanically too weak for glazing in the unfired state. The water in the glaze slip absorbed by the body (which is highly porous before firing) would cause the pieces to collapse. Only thick-walled ware can be coated with glaze in the unfired state.

Volatile media, such as alcohol, have not proved successful for applying glaze to pottery.

The density of the glaze slip depends on the degree of porosity of the body; the higher the porosity, the lower must be the density of the glaze slip. Non-porous bodies, such as bone china, require a glaze slip of high density.

The glaze slip should have good flow, but at the same time a high yield value is required to avoid running down verticle surfaces. Control of grain size distribution and additions of organic water-soluble polymers help in this direction. These polymers stabilize the glaze slip and act as binders to improve the hardness and abrasion resistance of the coated pieces before firing. Natural polymers, such as gums, celluloses and alginates, have largely been replaced by synthetic products, i.e. polyvinyl alcohol, and especially poly-acrylates, available as ammonia (NH_4) and sodium (Na) salts. Poly-acrylates can be tailor-made for specific purposes, either as low molecular weight deflocculants or as high molecular weight flocculants, to impart superior viscosity stability to the glaze slip on storage. As a result of their biological as well as chemical inertness, they are more resistant to bacteriological degradation than are naturally occurring polymers (Alston, 1974).

9.1. *Traditional Methods*

The glaze can be applied by brushing (only done by studio potters), pouring, dipping and spraying (aerographing).

Dipping, which used to be the most common method with industrially made pottery, is a skilful operation, especially with bodies of low or practically no apparent porosity.

The glaze slip is kept in open tanks. Constant recirculation which causes the slip to pass through sieves and over magnets guarantees freedom from dust, dirt, or other extraneous matter. The "dipper" places his ware in the glaze slip and immediately withdraws it, ensuring that the piece has been fully covered.

With porous pieces the water from the glaze slip is absorbed by the pores of the body, a solid layer of even thickness of glaze being formed on the surface of the article. The pores of the body are so small (less than 0.3μ) that they are not filled with solid glaze particles (coarser than 1μ).

In the case of non-porous pieces the water from the glaze cannot be absorbed by the body. Therefore the glaze coat stays wet after dipping. It depends on the skill of the dipper to shake and turn the article after immersion in such a way as to obtain an even coat.

9.2. *Advanced Methods*

9.2.1. *Automated Conventional Glazing*

The difficulty in obtaining an even coat of glaze on a non-porous body by dipping has led to the development of *spraying*. If done by hand, spraying requires skill and experience, as does dipping; moreover, hand spraying is too slow for modern production, but it can easily be mechanized and has, therefore, largely replaced dipping, especially as far as flatware is concerned. The articles are coated with glaze slip from spray guns (see Fig. 7.2).

Whereas dipping gives a solid, dense layer of glaze, spraying produces a more open, more porous coating on the articles; this causes a "rippling" effect. With the highly fusible glazes for bone china, vitrified hotelware and earthenware, the "ripples" are smoothed out during the "glost" (glazing) fire. However, with the much more viscous hard porcelain glaze the rippling effect is still vislble after firing; the high porosity (over 30 per cent) of the hard porcelain body, arising from its low "biscuit" fire (900–1000°C), is a contributary cause for the undesirable rippling effect.

Therefore, special glazing machines have been developed for hard porcelain. They have proved successful with cups and similar hollow ware whereby the inside of the vessel is coated with glaze slip by a fountain action and the outside by a waterfall action. Dipping machines constructed for hard

FIG. 7.2.   In-line flatware glazing machine. By courtesy of Service (Engineers) Ltd.

porcelain flatware had less success. For this reason the possibilities of spraying hard porcelain plates were again investigated. Through improved nozzle arrangements on the one hand, and introducing special organic compounds into the glaze slip (enhancing its flow properties when applied) on the other hand, a more solid layer of glaze and hence a satisfactory fired appearance was achieved, even with plates sprayed in the unfired state (Pfuhl, 1977).

These expedients may not always be sufficient to overcome the rippling effect; if so, the composition of the glaze has to be altered, making it more fusible.

The spraying machines for hard porcelain are incorporated in fully automated conveyor belt systems, comprising automatic brushing, stamping (fixing the manufacturer's trade mark), centring, removal of glaze from the foot of the plate, etc.

The requirements for mechanically spraying earthenware are opposite to those for hard porcelain. A light, powdery glaze coat gives better fired results than does a solid one. The quality of spray glazed ware has been found equal to or better than that obtained by hand dipping. The unit cost of production has been found to be substantially lower than that for hand dipping (Whitmore, 1974). The control of the following factors has ensured the required standard of ware:

(a) Density of glaze slip.
(b) Fluidity of glaze slip.
(c) Preheat temperature.
(d) Glaze and air pressure to the spray guns.
(e) Position of guns.

Waste of glaze used to be a serious problem with spraying. A great deal of engineering skill has been devoted to constructing devices for collecting over-spray by "labyrinth" filtering and recirculating solidified particles, reconstituted into glaze slip.

Spraying, whilst entirely satisfactory for general non-porous flatware, has not always proved suitable for hollow ware. The methods used for hard porcelain, viz. the fountain and waterfall systems, have been adapted and successfully applied to earthenware and non-porous biscuit ware, again incorporating complete automation (Hathersall, 1984).

Glazing by *disk centrifuge*, an American innovation (Murphy, 1983) used for tiles, has resulted in improved production efficiency and reduced cost.

9.2.2. Electrostatic Glazing

The electrostatic method of providing a coat of glaze was pioneered by a French company. It is characterized by a cloud of particles being produced with compressed air in a d.c. high tension field. The floating particles are electrically charged and migrate to the object to be coated (Heberlein, 1972).

Electrostatic glazing is particularly suitable for glazes on earthenware, vitrified hotel ware, bone china and stoneware bodies. (Adjustments in the composition of the less fusible hard porcelain glazes may be necessary.) Glaze consumption and the amount of rejects are reduced; fired glaze appearance is claimed to be superior to that of conventional glazing. The method is easily integrated into a fully automatic production line. Output is dramatically increased, compared with traditional glazing, since conveyor speeds of up to 20 m/minute are possible (Heberlein, 1976). Practical details of engineering requirements have been given (Lambert, 1974).

9.2.3. Fluidized Bed Dry Glazing

This process (see Figs 7.3a and 7.3b consists of the following steps (Roberts 1974):

1. Passing air through the dry powder of particles of glaze materials to which had been added a small proportion of resinous particles. The rate of air flow is such that the material suspended in the air stream acts as a fluid. As the stream of air passes upwards through the powder, friction produces a pressure drop which increases with the velocity of air flow. As the velocity increases, a point is reached when the pressure drop equals the sum of the weight of the bed per unit of cross-sectional area plus the friction of the bed against the walls. The bed expands and the particles forming this *fluidized* bed vibrate locally in a semi-stable, more open arrangement.
2. Preheating the ware to be glazed to between 100 and 200°C.
3. Bringing the heated articles into contact with the fluidized bed and keeping them there long enough for the resin, added to the glaze powder, to soften and to cause a layer of glaze material to adhere to the surface of the articles. After immersion in the bed, the specimens are placed on "pins" (see Glossary), ready for firing.
4. Firing the coated pieces to remove the resin and to mature the glaze in the normal way.

The advantages of dry glazing, using a fluidized bed are:

(a) Elimination of rheological problems arising from aqueous glaze suspensions.
(b) Avoidance of drying and hence saving costs.
(c) Uniform thickness of glaze layer, independent of porosity of the piece to be glazed.
(d) Full automation of the glazing process.

Experimental work highlighted the importance of grain size distribution. With the normal glaze fineness of 80 per cent less than 10 microns the process

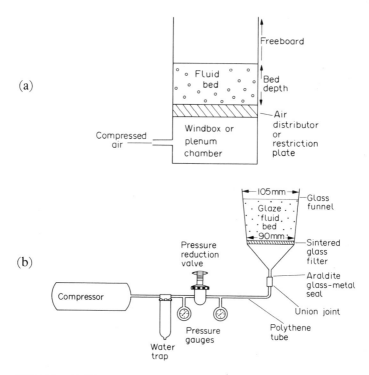

FIG. 7.3.   (a) Diagram showing the formation of a fluid bed. (b) Diagram of the fluidized bed apparatus. (After Roberts (1974), p. 48.) Reproduced by permission of The Institute of Ceramics.

proved to be very dusty. Much coarser material was required. The most suitable range was between BSS 120 and 200, the glaze consisting of 100 per cent frit.

The main problems were "crawling" (see section 5.2.2., p. 136) and lack of adhesion of the glaze to the ware in the early stages of firing, after the resin had been eliminated. The poor adhesion arose from the coarseness of the particles and could be overcome by applying a low-temperature "bridging" flux.

The process was not affected by the porosity of the article to be glazed, porous and vitreous specimens performing equally well.

It was concluded that dry glazing in a fluidized bed was ceramically feasible. Practical problems, such as the handling of pieces at temperatures of 200°C, would still have to be solved.

10. Drying and Finishing of Coated Ware

10.1 *Drying*

The procedure of drying and finishing is somewhat different for very porous and vitrified ware. Pieces of high apparent porosity lose their wet appearance a few seconds after the glaze is applied, although the solid coating obtained (as well as the body) retains appreciable moisture. Even so, in many cases such ware is not further dried before being put into the kiln, the very first stage of the glaze firing acting as dryer. The ware has, of course, a chance of at least partly drying before it passes into the kiln.

Drying of non-porous or slightly porous ware is necessary as the water from the glaze slip cannot be absorbed by the body. If the pieces were not dried they would remain wet and could not be handled. The drying of the glaze coat is not affected by the "memory of clay", as with the drying of the body, because the coating is thin and contains only little clay. The glaze coat can, therefore, be dried very much more quickly. Infrared dryers provide completely dry pieces in a matter of a few mintures.

In some aerographing and dipping machines the ware is heated before the glaze is applied. This facilitates subsequent drying.

10.2. *Finishing*

The finishing operation after the application of glaze is referred to as "retouching". The dipper has to hold the coated piece somewhere when he removes it from the glaze tank. Where he touches the piece with his fingers there are bare patches and glaze has to be applied to them afterwards. This is done by brush after drying; with high-class ware two "retouchers" are required per dipper. The retoucher not only applies glaze slip at spots where there is none but also removes dry glaze powder from the piece where there is an excess of glaze coating as in running marks. Their avoidance depends, to some extent, on the skill of the dipper in the first place. During firing the glaze becomes a viscous liquid which solidifies again on cooling. Therefore glaze has to be removed from the piece where it is in direct contact with the refractory on which it is placed during the firing.

With spraying and automatic dipping machines there are no bare patches or run marks. All the retouching necessary is confined to the removal of glaze from the feet of ware. This too is done automatically. In the case of porous pieces glaze is removed from the feet of ware by an endless rubber belt saturated with water. With the latest machines the belts vibrate, causing the glaze to be sponged away and the ware to be moved along the belt without manual help. An effective way of foot sponging is provided by two endless sponge belts, side by side, the one moving at twice the speed of the other; this causes the articles to move swiftly along the two belts, at the same time revolving around their own axes.

11. Firing of Glaze

The bulk of pottery produced in Britain and North America, e.g. earthenware, vitreous china, and bone china, is first fired biscuit to a high temperature, then coated with glaze, and the glaze matured by firing to a lower temperature. Hard porcelain, the staple type of pottery made in continental Europe (and some common pottery made anywhere), is fired to a low temperature first and glaze and body matured together at a high temperature. The kiln in which glaze is matured is termed "glost" kiln as opposed to the "biscuit" kiln. Similar types of kilns and kiln furniture are used, on the whole, for both biscuit and glost kilns, depending on temperature. Ware coated with glaze in the unfired state requires only one firing, body and glaze being matured at the same time, which considerably reduces production costs (see Chapter 6, section 7, "Once Firing", p. 121).

With low-temperature glazes, containing the oxides of lead, sodium and boron, there is the problem of volatilization. The loss of these volatile substances may cause the fired glaze to appear dull. Moreover, these volatiles attack the kiln refractories and cause actual glaze layers being formed on the roof of the kiln and the undersides of the shelves. This is not necessarily objectionable, in fact it may be desirable, as the glazed surfaces of the undersides of refractory shelves prevent bits of refractory material (which may have become loose during a succession of firings) from falling onto the ware, causing "kiln dirt". The undersides of the shelves are often purposely coated with glaze to prevent the ware from being spoilt by kiln dirt, and also to create a saturated atmosphere to avoid a dull glaze surface on the ware.

In the course of time there may have occurred an excess build-up of glaze on the roof bricks of the kiln so that "droppers" are formed. These consist of globules of glaze, up to 10 mm in diameter, liable to drop onto the ware which is thus spoilt. In bad cases portions of refractory may flake off and badly mar the ware.

Zircon ($ZrO_2.SiO_2$) and Zirconia (ZrO_2) are inert to glaze vapours and are therefore particularly suitable as refractory material for firing glazed ware. However, zirconium compounds, especially the oxide, are relatively expensive and, for this reason, cheaper and more conventional types of refractories are used, but coated with zircon.

One way of reducing if not eliminating the deleterious effects of glaze vapours is to cause sufficient air movement (draught) to move them out of the kiln. In tunnel kilns this is done by "contravecs".

Hollow ware is fired directly on the refractory shelves of the kiln (intermittent or tunnel), the pieces normally standing on their feet from which glaze has been removed.

Plates and saucers are fired as closely together as possible in special (open) setters called "cranks" (see Fig. 6.3b). In an attempt to reduce the amount of heat absorbing refractories to a minimum the method of placing flatware, known as "dottling", has been developed; in this case the plate is supported on its underside by three refractory pins. Sometimes flatware is set on edge, a method termed "rearing". This gives a better glaze surface because at peak temperatures the glaze is able to flow down a verticle surface.

Dottling and rearing have the advantage of allowing the plates, saucers, or dishes to have glazed feet since they do not rest on their feet. However, the backs of plates show pin marks.

Mass-produced plates, cups, and hollow ware of uniform height are glost fired in multi-passage kilns where refractories are eliminated (see Chapter 6, section 4.2.2., p. 111).

After the glost fire, pin marks (in the case of dottled or reared ware) and kiln dirt caused by bits of refractory dropped on to the glaze during firing are removed by polishing. This leaves marks on the glaze which look like devitrified patches. High-class ware is, therefore, reglazed (in such cases), usually by spraying the affected portion, and refired.

Fast firing of glaze (i.e. 2 hours) poses certain problems. Broadly speaking, with conventional firing the more fusible the glaze at the maximum temperature laid down, the more brilliant it will be and the more it will be inclined to self-heal imperfections. With fast firing the time allowed may not be long enough for this to happen. The expedient of raising the maximum firing temperature, as in the case of the biscuiting of the body (see Chapter 6, "Firing", section 1.2.4, p. 100), also works for glazes. However, increasing the firing temperature would defeat one of the objectives of fast firing, viz. reducing fuel consumption. Using completely fritted glazes helps as they are more fusible, their constituents having been prereacted. There is a trend for closer control of application techniques which should ensure that glaze imperfections are absent before firing.

12. The Properties of Glaze

12.1. *Has the Computer made Experimental Work Superfluous?*

The computer has enabled the research worker to predict a number of glaze properties (such as specific gravity, refractive index, Young's modulus, etc.) from its chemical composition. Complicated formulae have been worked out, analyzing and optimizing glaze properties so that experimental investigation would appear to have been made superfluous. The potential of the computer aid cannot be denied, but it is not a substitute for certain—visual—aspects of experimentation. The smaller manufacturers and studio potters do not usually have a computer at their disposal and thus have to rely on trials.

The dedicated research chemist will be able to prepare a few hundred glaze trials a day by prudent organization. It is beyond the scope of this textbook to describe the method of efficient glaze experimentation in detail; suffice it to mention that only a few (say four or eight) base glaze slips of extreme compositions need to be ground. Innumerable permutations can be produced in series by just volumetric mixing. A few small body tiles (30 × 30 mm) are dipped in each glaze and sets of the whole range fired at different positions in a production kiln. The effect of each serial replacement (varying only one oxide in the molecular formula at a time) can be seen at a glance and the general appearance evaluated, ranging from a matt surface to high brilliancy and gloss, from smoothness to egg-shell finish, from "crazing" to "chittering".

12.2 *Glaze Properties of Importance to the Manufacturer*

The most important glaze characterstic, as far as the manufacturer is concerned, is *thermal expansion*, which has already been discussed (see Section 3, "Body-Glaze Relationships", p. 126). Another aspect, highly significant in production, is *bubble formation*. Gas bubbles are invariably present in glazes; they are unavoidable. Their number, size and position relative to the surface and the interior determine the visual quality of the glaze. A poor glaze surface is due to an unacceptable amount of bubble. Excess bubble formation may manifest itself in the following ways:

(a) The bubbles are uniformly distributed throughout the thickness of the glaze layer—dull surface.
(b) The bubbles have burst and healed—crater surface.
(c) The bubbles are under the surface—orange peel surface.

The mechanism of bubble formation has been studied with the aid of a hot stage microscope: as the glaze approaches its maturing temperature, there are a great number of bubbles throughout the layer. Gradually, as the maximum temperature is reached, the larger bubbles have swallowed up the smaller ones so that overall there are fewer bubbles. A few large bubbles provide a better glaze surface than do many small ones.

Bubble formation is also governed by the underlying substrate, i.e. the body (Dinsdale, 1986). (The interfacial layer between body and glaze may also play a part.) Bubbles tend to attach themselves to relatively large particles (above 20 microns) of crystalline free silica protruding from the body surface. In the course of studying the influence of the body on glaze surface quality, an identical glaze showed very many bubbles if applied to an earthenware body (having a large amount of silica crystals), few on bone china (with possible little free silica, not taken up in the formation of anorthite and glass) and none on a pure alumina ceramic. There was a marked progressive improvement of the glaze surface.

[NB: The glaze characteristics of importance to the user are discussed in Chapter 10, "The Properties of Pottery".]

Further Reading

General: Parmelee (1951); **Taylor and Bull (1986)**; Singer and Singer (1963, pp. 526–660, 777–797, 961–966).
Origin of glass and glazes: Flinders-Petrie (1924–5).
Structure of glazes: Moore (1956); Singer and German (1960); Smith (1953).
The glassy state: Knott (1983).
Properties of glass: Moray (1954).
Low-temperature glazes: Singer (1954).
Glaze-body reactions: Roberts and Marshall (1970); Kerstan (1982).
Grinding of glazes: Slinn and Rodgers (1980).
Automatic glazing machines:
   Spraying: Whitmore (1974); Pfuhl (1977).
   Other: Hathersall (1984); Murphy (1983).
Electrostatic glazing: Heberlein (1972, 1976); Lambert (1974).
Fluidized bed dry glazing: Roberts (1974).
Optimization of glaze properties by computer: West and Gerow (1971).
Glaze defects: Kure (1957).
Durability of glazes: Mellor (1935).

8

Decoration

Decoration of pottery has played a part in pottery ever since the first pots were made. The early potter who moulded his vessel with the aid of baskets saw that the basket weave was imprinted on his pot. Having found this first type of decoration pleasing he continued to engrave the patterns derived from the baskets and similar designs on his pots long after he had discarded the baskets as an aid to potting and gone over to other methods, such as throwing.

Glazing can also be regarded as a form of decoration. Furthermore, polishing of pottery which was done at various stages of its development produced original decorative effects.

The most striking decoration is achieved by colour and this field of decoration concerns us in this chapter. Both bodies and glazes can be coloured. However, in industrially made pottery, colour is more often used in the form of decorative patterns. These are applied either on the biscuit (under-glaze) or the glost-fired piece (on-glaze). The latter application requires another heating process, viz. the decorating firing, rarely exceeding 800°C.

Such decoration is really an ancillary process in the production of pottery although some people regard it as the most interesting and important stage. There is a clear division between the potter and the decorator who applies patterns on the finished ware. In the view of some potters on-glaze decoration (and to some extent under-glaze decoration) does not belong to pottery. They think it is superfluous and regard it as gilding the lily. Yet white pottery is rich in examples of exquisite hand painting or printing.

1. Ceramic Colours

Colour in pottery is almost invariably provided by certain metallic oxides. The metallic oxides are seldom used alone but are usually mixed with either a filler, if used at high temperatures, or a flux, if used at low temperatures. A filler, such as alumina, quartz, feldspar, china clay, actual body, etc., "stretches" the colour and at the same time makes the metal oxide more refractory. Some colouring oxides act as strong fluxes in themselves and for this reason refractory fillers are added to them even for low-temperature application. Most on-glaze colours require fluxes. The addition of filler or flux not only helps to reduce the price of the stain, the individual oxide colours being usually expensive, but also ensures greater uniformity, stability, and consistency.

The final stains produced, ready for application, are not necessarily oxides. The majority of ceramic stains are silicates but there are also aluminates, phosphates, uranates, etc. Several ceramic stains have spinel structure (see Fig. 8.1). Normal spinels consist of the general formula AB_2O_4, where A stands for a divalent cation and B for a trivalent cation, each A atom being co-ordinated with *four* oxygen atoms (i.e. in tetrahedral co-ordination) whilst each B atom is co-ordinated with *six* oxygen atoms (i.e. in octahedral co-ordination). Spinels are the most stable of all ceramic stains, at least in bodies. Because of their stability there is no radical colour change when they are diluted with filler or flux.

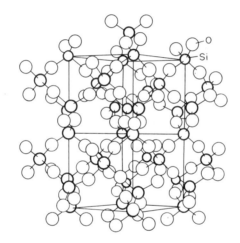

FIG. 8.1.   The structure of normal spinel (After Worrall (1986), p. 216.) Reproduced by permission of Elsevier Applied Science Publishers Ltd.

Temperatures in the region of 1800°C are required to form spinels by sintering. Mineralizers, such as boric oxide, help to reduce the temperature necessary for these solid–solid reactions. Many ceramic stains are formed in the presence of a liquid phase.

Ceramic colours are mainly used in the form of powdered stains which, in the cold state (state of application), are inert; they can also be applied as solutions, but this method is now comparatively rare.

A number of the colouring metal oxides are only stable within certain temperature ranges. The palette of colours available is far greater for the low-temperature ranges. Colour tone is influenced by the following factors:

(a) The temperature of firing.
(b) The atmosphere of firing.

(c) The calcination temperature and fineness of the colour.
(d) The type of filler or flux used.
(e) The compound used for introducing the metal oxide concerned.
(f) The type of application, e.g. whether used to stain the body or glaze (and in such cases the composition of body or glaze itself) or applied under-glaze or on-glaze.

2. Preparation of Stains

The fine powders of colouring compound, filler, or flux are *blended*; this is usually done by water-grinding followed by drying. If the stain composition contains water-soluble salts it is dissolved in water and mixed with the dry components to a paste, which is dried and pulverized so that the soluble salts which migrated to the surface are redistributed; thus local concentration of colour is eliminated.

Blending is followed by *calcination*, which develops and stabilizes the colour. The temperature of calcining is, as a rule, higher than the maximum temperature at which the stain is applied. The finely dispersed colour mixture is fired (usually in saggars) to a closely controlled schedule.

The hard cakes formed during calcination are *crushed*, *sieved*, and *water-ground* to great fineness. Any soluble salts liberated during grinding—causing "halos" on the ware or giving the colour a smudgy appearance—are removed by *washing* with boiling water. The leached stain is *dried*, *pulverized*, and *screened* through the finest mesh commercially available.

With certain difficult colours, e.g. the well-known cobalt blues, the prepared stain is sometimes treated like a raw colouring oxide and blended with a further filler, recalcined, crushed, ground, washed, and screened. This is done to make it yet more foolproof; such an expedient was employed in the sixteenth century in the making of Medici porcelain (Chapter 9, section 2.2.2., p. 181).

3. Colouring Agents

Colouring agents are listed in Table 8.1.

As will be seen from Table 8.1, the following elements figure most prominently as ceramic colouring agents: *chromium, cobalt, iron, uranium, copper* and *gold*. Considerable use is also made now of *vanadium* and *zirconium*.

Describing the basic principle of how colour is produced in ceramics, Singer and Singer (1963) conclude that "the colour of an atom in combination depends on its environment". This is particularly striking with *chromium*. Compounds of this element normally give greens but, in combination with tin and alumina, pinks and reds are produced. Provided the temperature is high enough (approaching 1800°C) to allow solid solution to be developed to the full, additions well below 1 per cent of chromia to alumina (which is pure white) produce deep ruby shades. Small amounts of chromium added to porcelain glazes containing magnesia give pleasing brown tints, whereas in glazes without magnesia a pure green is produced. (Stains based on chromia have been made which, introduced into porcelain glazes, change colour if exposed to different light sources. Thus a pure green in daylight can be turned into a mushroom grey in electric light; likewise a mushroom grey in daylight is transformed into pink in electric light.)

The influence of the environment is far less pronounced with *cobalt*, which, like most other colouring elements, undergoes changes in shade. However, it does not give completely different colours, like chromium, if exposed to different environments. It is famous for its deep, rich blues. Mixed with chromium it gives blue-greens, and added to mixtures of chromium and iron it produces a black colour. Phosphate of cobalt is pink and if cobalt oxide is introduced into bone china, the only phosphatic body amongst pottery or other ceramics, the resulting stain is pink to lilac. However, this colour tone is unstable and is destroyed by the glaze. Cobalt shares with chromium the ability to withstand high temperatures and is most effectively used under-glaze. It is often applied on-glaze and refired at glost temperature, even with hard porcelain (1400°C), giving the well-known beautiful tints of royal or mazarin blue.

Iron, also stable up to high temperatures, again covers a wide range of tints, largely depending on atmosphere, covering brown, yellow, orange, off-red, purple, black, and, in the famous Chinese Celadon glazes, green.

At one time compounds of *uranium* were extensively used as colouring agents for yellow, orange, and red (in oxidizing atmospheres), as well as black (in reducing atmospheres). The supply of uranium to the pottery industry was completely cut off when it was first employed for atomic energy. However, depleted uranium oxide (U_3O_8) is again offered to the ceramic industry. It is greatly influenced by atmosphere.

Copper, although rarely stable at high temperatures, is similar to chromium and iron in that it produces different colours under different conditions. Normally copper is used for green shades at low temperatures. In highly alkaline low-temperature glazes it turns into a beautiful turquoise (marine blue). The magnificent reds, originally made in ancient China and reproduced in Europe since the nineteenth century, known as *rouge flambé*, *sang-de-bœuf*, etc., originate from copper and are only obtained under reducing atmosphere introduced at the right stage. The state in which the copper is present in *rouge flambé* is still disputed, colloidal metallic copper and the suboxide of copper having been postulated.

Gold has been in use for making the family of pink to purple colours since the nineteenth century. It is introduced as gold trichloride and in the fired state is thought to be present as colloidal gold, the actual shade of colour depending on the particle size.

TABLE 8.1.　*Colouring Agents*

Colour	Colouring agent	Temperature range, stability, etc.
White	Tin oxide, suitably fluxed, e.g. lead silicate	Decorating and glazes
	Zirconia	Replacing tin oxide as much cheaper
Black	Mixtures of oxides of iron, cobalt, chrome, manganese, and occasionally nickel	All temperatures
	Uranium oxide (U_3O_8) depleted	High temperature in reducing conditions only (hard porcelain)
	Iridium oxide	Very expensive
Grey	Diluted black agents, also platinum	All temperatures
Yellow	Tin oxide plus vanadium compounds	Not stable beyond 1100°C
	Zirconium vanadates	Not stable beyond 1100°C
	Oxides of antimony, cadmium, praseodynium, titanium, usually with iron, lead chromate	Not stable beyond 1100°C
	Uranium oxide depleted	In oxiding atmospheres only
Blue	Compounds of cobalt	All ranges of blue including deep shades. All temperatures
	Vanadium-zirconium silicates	Turquoise shades only; up to 1280°C More stable than cobalt compounds
	(Nickel oxide	Low-temperature glazes only)
	(Titanium oxide	Under certain conditions only, very unreliable)
Green	Chromium sesquioxide and other chromium compounds	Stable up to high temperatures; colour tone largely dependent on filler, glaze, or body; magnesia, high content of alumina, manganese to be avoided
	Copper compounds	Up to 900°C only; must be fired oxidizingly; high alkali glazes to be avoided
	Vanadium silicates with zirconia (also tin oxide)	Up to 1250°C
Brown	Iron compounds (spinels) possibly with titania and chromia	All temperatures; atmosphere dependent

Gold (and platinum) is used for metallic decoration of pottery. Only the precious metals are suitable for this purpose as base metals would not resist oxidation during the decoration firing. "Liquid" or "bright" gold (the gold being present in the form of a soluble compound) in an oily medium provides an extremely thin coating, about 0.1 μ. A thicker, more lasting and, incidentally, more pleasing coat is obtained by so-called "burnished" gold made with mixtures incorporating powdered gold or gold amalgamated with mercury; this appears dull after firing and has to be burnished (see section 4.4.9, "Burnishing", p. 163).

As a result of the discovery in the United States, in 1948, that certain colouring ions could be "latched" into the *zircon* (zirconium silicate, $ZrSiO_4$) lattice, a new field of colouring agents was opened up. The first successful

TABLE 8.1.   *continued*

Colour	Colouring agent	Temperature range, stability, etc.
Brown- *cont.*	Manganese compounds Chromium sesquioxide	Not stable at high temperatures For high magnesia porcelain glazes only

Red	Ferric oxide plus zinc oxide	Stable at low temperatures only; only
		  possible with correct calcination
		  temperature

	Chromium–tin compositions, lead chromate	Stable at low temperatures only
	Chromium–alumina compositions (plus boric acid)	For high temperatures
	Cadmium–selenium compositions	For low temperatures only
	Uranium oxide depleted	Difficult, usually only orange tints, in oxidizing conditions only

	Colloidal metallic copper (possibly
	  copper suboxide)
		Only possible under most carefully
		  controlled reducing conditions; the
		  only red possible at hard
		  porcelain temperature

Pink	Diluted reds, particularly chromium–tin bleached by alkalis	
	Manganese–alumina compositions	Suitable for staining bodies, e.g. hard porcelain and bone china

	Gold chloride solution	For low temperatures. Also in hard
		  porcelain and bone china bodies
		  (giving beautiful tints, but very
		  expensive)

	Cobalt compounds	For bone china body only
Purple	Gold chloride with tin chloride	Generally for low temperatures only

	Cadmium–sulphur–selenium
	  compositions
		Increasing amounts of selenium
		  giving maroon tints

Gold	Metallic gold	For low temperatures only

Silver	Metallic platinum, also palladium
	(Metallic silver
		For low temperatures only
		Seldom used because of tarnishing,
		  particularly under the influence of
		  sulphur fumes present in most
		  industrial areas)

colouring ions were provided by *vanadium* compounds with the addition of an alkali fluoride mineralizer. A typical stain had the following composition (Batchelor, 1974):

	wt%
Zirconia	62
Silica	30
Ammonium vanadate	5
Sodium fluoride	3

Calcined at 850–1000°C, this composition gave a blue colour. Greens and yellows could be produced by altering the percentage of the vanadium compound. Pinks were obtained by combining zircon and iron compounds.

These zircon-based stains are highly stable up to 1100°C, but not in reducing atmospheres; they are thus unsuitable for hard porcelain bodies, glazes and under-glaze decoration.

4. Application of Ceramic Colours

4.1. *Coloured Body*

We usually regard white pottery without patterns of colour as "undecorated". If colour is present either as design on the piece or in the body or glaze, then the piece is called "decorated". Considered in this light the first pottery that was made was decorated. Early man made his pots from unrefined clay. This invariably contained a high amount of iron oxide which acted as a colouring agent, giving a variety of brown, red, or yellow colours, or, if fired in a reducing atmosphere, very dark tints.

Natural clays do not usually contain any of the other ceramic colouring agents; if green, blue, grey, or pink bodies are required the appropriate calcined oxide stains are added to a white body. They almost invariably contain a high percentage of refractory filler for reasons already stated (see section 1, p. 150).

The amount of prepared stain (including filler) does not usually exceed 5 per cent. Very attractive delicate hues are obtained with less than 1 per cent.

As the housewife makes her washing look whiter by "blueing" it, so does the potter add a "blueing" agent in the form of a cobalt compound to the body. Less than 0.01 per cent cobalt, often introduced as a solution to get completely even dispersion, has quite a remarkable effect, in making the ware look "whiter".

The finely ground stains are incorporated in the body in the grinding cylinder, the clay blunger, or the mixing ark.

4.2. *Engobing and Slip Decoration*

In his endeavour for refinement the potter tries to make his dark-coloured clay or pottery body appear white. He can do this by an opaque white glaze but it is easier and more foolproof to cover the dark-coloured body with a thin coating of white body. This coating applied at some stage prior to glazing, is called *engobe*.

The reader may ask: why not use the white body in the first place? The answer is: the white body would be far more expensive and its plasticity would be far inferior so that the potter would not be able to use it for the particular shapes he would want to make; in any case it would take him much longer and he would have more failures.

The thermal expansion and shrinkage of the engobe and body proper have to be similar.

The engobe can be coloured by a stain incorporated in it in the same way as in a white body.

One of the most attractive methods of decoration for which old-English peasant pottery is famous, now being perpetuated by studio potters, is slip decoration, sometimes referred to as "slip trailing". Here a second engobe of contrasting colour is applied on the first after the latter has partially dried. The second engobe provides the pattern and it is normal practice to apply the "trailing" (second) engobe, as a coloured slip, on the white (first) engobe. Whilst the first engobe is applied by dipping, spraying, or brushing, the "slip" (second engobe) is brushed on freehand or squeezed through a perforated stencil; more often it is applied by a bulb and quill in the same way as the pastry cook decorates a cake. Much of this slip trailing is done while the piece is rotating on the wheel. The fact that the application is liquid (only water suspensions being used) offers wonderful possibilities of decorative effects, e.g. by simply connecting a series of horizontal bands (produced while the wheel is spinning) with vertical lines. People without any natural artistic ability, complete novices in the art of potting, will be delighted to discover that by relatively crude means they are able to produce what seems—at least to them—pleasing and quite unexpected designs. Slip decoration is far from precise but therein lies its special charm.

A more refined way of slip decoration is *pâte sur pâte* decoration on porcelain: a low relief of successive layers of white porcelain body is built up on a coloured base providing an original effect due to the translucency of porcelain.

The opposite to slip trailing is *sgraffito*. With this method the piece is covered with the engobe and when this is nearly dry, patterns are produced by scraping portions of engobe away exposing the underlying body. Sgraffito and slip trailing can be combined.

Sometimes a coloured body is incorporated in white pottery. Hollow ware, coloured outside and white inside, is simply made by pouring coloured casting slip into the mould, emptying it very shortly afterwards, followed by immediately pouring white casting slip in the mould on to the very thin cast of coloured slip. The white slip is poured off in the normal way when the necessary cast thickness is reached. Likewise, flatware, coloured on top and white underneath, is made by first jiggering a thin, coloured layer followed by jiggering the white layer immediately afterwards. If coloured rims are to be produced the centre portion of the coloured layer is simply cut out before the white portion is jiggered. The same effect can be produced by slip casting the rim before jiggering the white part. The slip casting is done by a special machine which confines the slip to the rim (Worcester Royal Porcelain Co., 1960). The great advantage of this method lies in the fact that there is no need for cutting out the centre portion as with the jiggering method.

4.3. *Coloured Glaze*

Glazes normally appear colourless on the body, although by themselves they are seldom without some tint. Lead glazes are slightly yellow and boron glazes slightly blue. The high-temperature glazes rich in alkali as used for hard porcelain are somewhat white opaque. To colour a glaze, metal oxides (other than the opacifying agents) may be added to it. These oxides give transparent glazes, creating an aesthetic unity with the body (Mettke, 1984). More often, prepared stains are incorporated in the glaze, producing an opaque appearance.

Stains for bodies can also be used in glazes. Most colouring agents and stains act as fluxes in the glaze, causing it to soften. This softening may help to bring about an additional artistic effect. Where the glaze layer is thin, e.g. at the edges, the colour is much lighter. On ware with relief modelling the coloured glaze has a tendency to run off the sharp edges and raised portions of the relief and be collected in thick pools in the depressions of the relief. The varying thickness of glaze, light in thin layers, dark in thick layers, enhances the aesthetic appeal of the piece.

If the inside of hollow ware has to be white and the outside coloured, the white glaze is first poured into the vessel. When this glaze slip has reached the rim of the piece it is immediately poured out again, the piece retouched and carefully lowered into the tank containing the coloured glaze until the glaze has reached the rim of the piece, when it is quickly removed from the glaze tank.

Very interesting effects can be achieved by so-called "running glazes". They are mostly applied to vases. Blobs of stain are put under the top or on the shoulder of the vase already carrying a coloured glaze. The stains contain strong fluxes so that at the temperature at which they are fired they become liquid and run down to the bottom of the piece. In doing so they not only merge with the coloured glaze on the ground but with each other so that each piece is different. The colours run beyond the foot of the piece on to the refractory support and this involves a good deal of grinding and polishing.

Other decorative glaze effects where colour may be involved have already been referred to in the section on special glazes (Chapter 7, section 5, p. 133).

Fast firing may cause problems with coloured glazes. There may not be time for the necessary reactions to take place. Normal stains may have to be replaced by "metal frits" in which the state of "prereaction" is at a more advanced stage, due to higher flux content.

Stains often influence the thermal expansion of the glaze, usually increasing it and thus leading to crazing; the measures indicated in Chapter 7, section 3.1 ("The Significance of Thermal Expansion") p. 126, would have to be applied.

4.4. *Traditional Methods of Under-glaze and On-glaze Decoration*

The most common form of colour decoration in whiteware (earthenware, bone china, hard porcelain, etc.) is by patterns to (a) the body before it is covered by glaze (under-glaze) or (b) more often on top of the glaze (on-glaze, or over-glaze in the United States). Firing on-glaze decoration to a low temperature (viz. 700–820°C) causes the colours to soften and unite with the glaze. Different on-glaze colours may require different firing temperatures; this implies that some pieces have to have more than one decorating fire. In rare cases the decoration is painted on the unfired glaze after dipping and drying, glaze and colour being matured at the same temperature, as in under-glaze decoration.

Organic media are used for applying under-glaze and on-glaze decoration by ceramic colours. Amongst the traditional ones are linseed oil, turpentine, resins, Canada balsam, gums, tars, etc. As these natural materials are likely to vary they are being largely replaced by synthetic media, mostly synthesized resins such as formaldehydes, alkyds, polystyrene, vinyl acrylic resins, stearates, etc.

4.4.1 *Freehand Painting*

The colours making up the design or pattern are applied by brush, although the technique is different from painting with oils on canvas (Fig. 8.2). For bands and lines the piece is centred on a horizontal wheel which the decorator slowly rotates with his left hand while the brush in his right hand touches the rotating plate, his right hand being kept still on a hand rest.

Apart from printing freehand painting used to be the only method of on- or under-glaze decoration and even today a few exclusive factories, notably the Nymphenburg Porcelain "Manufactory", employ no other method of decoration.

4.4.2 *Spraying*

This is done in the same way as the spraying of glaze and gives a more uniform result than brushing. The portions not to be covered by colour have to be screened by some kind of "resist".

Spraying under-glaze produces beautiful effects on hard porcelain, Danish porcelain being famous for this.

4.4.3. *Ground Laying*

Like spraying, this method provides a "ground", viz. wide bands, etc., not actual patterns. Ground laying is slow, requiring great skill, and is therefore

FIG. 8.2.   Hand painting of vase. By courtesy of Royal Worcester Spode Ltd.

very costly. It is only used for certain difficult on-glaze colours (which lack uniformity if applied by other methods) on expensive dinner ware in bone china. The parts to be left free from colour are covered with a resist and the remainder of the piece is painted with a drying oil. When the oil has dried to the right tackiness the piece is "bossed", viz. all brush marks on the drying oil are removed by striking with a thick pad. The powdered colour is then dusted on with a pad of cotton wool and the excess shaken off under an extractor hood. The piece is dried and the resist removed by washing in warm water.

4.4.4. Printing

Next to hand painting, printing, also referred to as "offset printing", first introduced in the mid-eighteenth century, produces the most beautiful and accurate on-glaze and especially under-glaze decoration. The design is engraved or etched on a copper plate. The engraving is filled with the oil-based colour and printed on tissue paper by passing copper plate and tissue paper through a heated press. The tissue paper is removed from the copper plate and then used as a transfer, being rubbed on to the piece coated with size, thereby fixing the design on to the piece. The tissue paper is then removed with the aid of a sponge. Where an exceptionally high standard is required this costly manual method as first introduced over 200 years ago is still used. Some ware produced at Worcester which is famous for this method is shown in Fig. 8.3).

4.4.5 Silk Screen Printing

This relatively new process is suitable for both mono- and poly-chrome decoration. It is a stencil operation in which the stencil is supported by a mesh fabric, usually silk cloth but also synthetic fibres or wire screen. The design is printed by drawing a rubber blade over the stencil which is filled with ceramic ink. The ink is forced through the open areas of the stencil and is deposited on the piece. The stencils are often made by photographic methods.

4.4.6. Lithographing

Lithographing, known as *Decalcomania* in the United States, is the most widely used method of on-glaze decoration; it is also suited to under-glaze application. the multicoloured patterns are supplied by the "litho" manufacturers in the form of large paper-backed sheets. These are cut up by the operative into easily manageable segments which are applied to the articles as with printing. The ware used to be painted with a thin layer of varnish before applying the transfers to provide a tacky surface.

The need for varnishing has been completely eliminated by the use of *water slide transfers*. Using the silk screen method or ordinary lithographic means, the pattern is printed by the litho manufacturer onto special lithography paper and is covered with a plastic medium. This cover-coat, containing the design, is in turn covered by a layer of water-soluble gum. The operative soaks the transfer in water and slides the cover-coat of the backing paper. Owing to the water-soluble gum, the cover-coat can easily be stuck to the articles without the need for rubbing and sponging.

The plastic cover-coat has a certain flexibility and ability to be stretched. It can, therefore, be applied to surfaces which are not flat and to ware,

FIG. 8.3. Vase showing the highly developed art of printing, late eighteenth-century Worcester. By courtesy of Royal Worcester Spode Ltd.

especially large dinner plates which, as occasionally happens in pottery production, are slightly over-sized. It is important to ensure that no air is trapped under the transfer. The slide-off transfer process has ousted the former lithographing method, involving varnishing, as it is much quicker and requires less skill.

4.4.7. Stamping

This method is applied to cheaper ware, since it lends itself to a high degree of mechanization. It supplies patterns by a rubber stamp, fixed to a small roller, which rolls round the article, alternately picking up and applying colour.

4.4.8. Acid Gold Etching

In this form of decoration, part of the glaze is removed by etching with hydrofluoric acid. This produces a raised pattern on a matt, receded background. A deeply cut copper engraving is filled with an acid-resisting substance (resist), such as beeswax, and transferred to printing paper which is applied to the glazed piece. The remaining areas not to be etched are covered with the resist so that only the engraved portions to be etched are left exposed. The article is immersed in hydrofluoric acid for a few minutes (the time of immersion depends on the concentration of the acid) and rinsed in water. The resist is removed and the gold applied to the engraved portion. Designs obtained by this process have a rich and, at the same time, elegant effect.

4.4.9. Burnishing

This process only concerns burnished gold, acid etched as well as normally applied, and platinum. When burnished gold leaves the decorating kiln it is matt and dull. Rubbing it with agate, haematite (bloodstones), or zircon sand makes it look brilliant (p. 154).

4.5. Mechanized and Automated Decoration

In the past, decoration by mechanical means had been the "Cinderella" among tableware processes. However, since the nineteen-sixties great progress has been made, especially in Britain, in the development of decorating machines.

4.5.1. Lining and Banding Machines

A number of lining and banding machines, including multicolour ones, have come on the market, incorporating self-centring and other sophisticated devices. These machines allowed much faster production than did manual banding (Ellis, 1973); therefore, special thermo-plastic applicators were developed which dried much faster than conventional media.

4.5.2. Offset Printing Machines

The first mechanized device for offset printing was the MURRAY CURVEX MACHINE, brought out in Britain in 1955. It consisted essentially of a convex gelatine pad which was lowered, under pressure, onto an inked engraving. This pad, while being deformed, picked up the inked trace from the engraving. It was brought into contact with the article so that the design was transferred to the surface of the article. The machine was semi-automatic and confined to single-colour printing.

A multicolour printing equipment was developed (also in Britain), known as "DEKRAM", which allowed different colours to be applied, one after the other, by a series of printing pads. Between each colour application a powdered substance, such as starch or methyl cellulose, was applied to the freshly printed colour to prevent it from being spoiled by the subsequent printing pad. This method has been superseded due to the introduction of thermoplastic inks (see below).

Although materially cheaper than the very time-consuming manual printing, machine printing remained unduly expensive in relation to the total cost of the article carrying the print. In the search for a more economic method of high-quality decoration, an assessment was made (in 1976) of the merits and drawbacks of the principal decorating methods, considered from the state of mechanization, as shown in Table 8.2 (Roberts, 1981/4). Some interesting facts emerge from this table:

Offset printing, the oldest decorating method (leaving aside hand painting), still offers the greatest promise

(a) as a method suitable for applying high-quality decoration directly onto an article, independent of complexity of shape,
(b) as having the potential of being developed into multicolour designs, and
(c) most importantly, as a process which lends itself to automation and high speed of production, hence low cost.

The development of automated offset printing has been described very fully by its pioneer (Roberts, 1981/4, 1981/5).

Gelatine vs. Silicon Rubber Pads

From a survey by the British Ceramic Research Association in the late nineteen-seventies, it emerged that the difficulties arising from printing machines, then in use, had their cause in the gelatine; it had poor mechanical strength, poor resistance to compression, variable hardness, depending on

TABLE 8.2. *The State of Mechanization in 1976 of the Main Tableware Decoration Processes (after Roberts, 1981/4). Reproduced by permission of Verlag Schmid GmbH*

Process	Advantages	Disadvantages
Transfers	Multicolour Halftones	Hand application Difficult to mechanize
Silk screen printing	Multicolour	Limited to flat and cylindrical shapes
Lining and banding	Mechanized	Limited to the edge only in the case of holloware
Stamping	Mechanized	Fairly crude designs
Offset printing	Printing directly onto complex shapes	Semi-automatic only Limited designs Monochrome

temperature, and other disadvantages which caused inconsistent quality of the print, requiring constant adjustment of the machine by the operator.

A replacement of gelatine as a pad material was obviously called for. Among the numerous materials tried, *silicone rubbers* proved most suitable. They were prepared from a basic polymer, polydimethylsiloxane, by using a cross-linking agent with the addition of a catalyst at room temperature. Rubbers cured at room temperatures proved too hard; they could be made softer, even than gelatine, by the addition of a silicone fluid during fabrication. Such silicone rubbers displayed the properties required, i.e. high deformability for low pressure, good elastic recovery and toughness.

The fact that silicone rubber pads can be manufactured that are not only soft but tough has focused attention on the possibility of printing directly onto the unfired body.

The silicone rubber system has the following advantages:

1. Improved printing consistency.
2. Longer life of printing pad, not affected by humidity and temperature.
3. Simplified printing operation, leading to a more intensive automation.
4. Promise of more consistent registration in multicolour definition.
5. Improved print definition and dimensional accuracy.

These advantages arise from the thermal stability of silicone rubber. Pads made of this material can, therefore, be used in conjunction with thermoplastic inks.

Cold vs. Thermoplastic Printing Inks

The ink used for gelatine printing proved unsuitable for silicone rubbers. The development of high-speed machines demanded fast drying inks. Thermoplastic inks, when heated to temperatures around 80°C, are fluid enough to wet the rubber at the design plate stage and "freeze" on printing. These thermoplastic inks thus prevent problems, associated with "wet-on-wet" printing when cold inks are used.

The use of these thermoplastic inks opened the way to multicolour printing which has enormously increased the potential of the offset printing process.

Metal vs. Photopolymer Design Plates

With the perfection of mechanized printing, a less expensive alternative to the hand engraved copper design plate was sought to reproduce halftone work. With hand engraving, a highly skilled craft, a variety of tones can be reproduced by varying the depth of the cut. Copper is rather soft and in order to provide a harder surface, the copper plate is chromium plated. However, even the chromium plated design plates are insufficiently durable for modern production runs. Longer life is obtained with plates, acid etched in hardened steel.

A further improvement is possible by using an inexpensive plate with a light-sensitive plastic working face (similar to that introduced in the letterpress printing industry), bonded to a metal base for rigidity. The principle on which the plate works is based on the fact that exposure to ultraviolet light polymerizes the plastic, causing it to become permanently hard. The unexposed areas can then be washed away with an appropriate (e.g. alcoholic) solvent, which has no effect on the hardened areas.

A photopolymer plate can be produced once a design is translated into a photographic film. The photopolymer, i.e. photosensitive nylon, gives an etch of excellent resolution, reproducing halftone work and line work accurately. The plates are made to engineering tolerances and are, therefore, ideally suited to automated processing. The photopolymer design plates have the advantage of low cost and high processing speed and, moreover, open up new fields in short-run production which would have been commercially impractical with the expensive hand-cut plate (Basnett, 1980).

4.6. Computerized Decoration

An Italian firm has put on the market a computer controlled painting machine "Plotex 3" (see Fig. 8.4). It is capable of painting with five different colours, of spraying, with or without stencils, and of engraving patterns into the dried glaze coat. The machine is controlled and programmed by a computer the modules of which reproduce the different movements. This decorating unit can be incorporated in all production lines (Luchs, 1985).

5. Fast Firing of In-glaze Decoration

Normal on-glaze decoration is fired as low as 750–820°C.

The application of decoration to the fired glaze, followed by fast firing to temperatures high enough for the colours to sink into the softening glaze, has already been dealt with under section 6.3.3, "Making Decoration Dish-

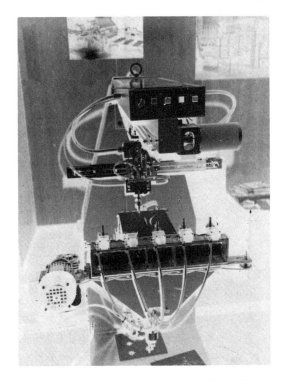

FIG. 8.4.   Computer-controlled painting machine "Plotex 3" by Palladini and Giovanardi. Reproduced by permission of Verlag Schmid GmbH.

washer-proof" in Chapter 6 (p. 120). This relatively new method, known as *"in-glaze"* decoration, is compared schematically in Fig. 8.5 with the traditional on-glaze and under-glaze decorating processes, showing the differences between hard porcelain on the one hand and bone china, vitrified hotelware, earthenware, etc., on the other hand. The purpose of rapid firing of in-glaze decoration is reiterated, viz. to reduce firing costs, and to make a larger palette of colours durable.

It is unlikely that the rapid decorating firing schedules of 60–90 minutes, as used for hard porcelain, will be applied to earthenware (Bull, 1982). To compensate for the short duration of firing, the maximum temperature has to be increased. The higher peak temperature causes a fault in the porous earthenware, known as *"spit-out"*; this consists of numerous small craters in the glaze, brought about by desorption of previously adsorbed water vapour.

The fast firing of decoration on bone china and other vitreous bodies has proved successful. The widest application is with hard porcelain; 80 per cent of tableware decoration on hard porcelain in Germany, was *fast* fired in 1985 (Pfaff, 1985).

A strongly oxidizing atmosphere is required in the early stages of fast firing in-glaze decoration (200–500°C) in order to burn out the resins. Butylmethacrylate polymers, generally used in ceramic transfers, have excellent burn-out properties (Bull, 1982).

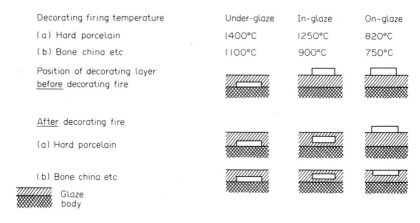

Decorating firing temperature	Under-glaze	In-glaze	On-glaze
(a) Hard porcelain	1400°C	1250°C	820°C
(b) Bone china etc.	1100°C	900°C	750°C

FIG. 8.5.   Comparison of fast fire in-glaze decoration with on-glaze and under-glaze decoration.

When the temperature is high enough for the fluxes to melt, the pigments become embedded in the glassy layer formed. The highly mobile ions of alkali, alkaline earths, also boron, lead and zinc oxides of the glassy layer, diffuse into the glaze. With hard porcelain the ion exchange thus brought about results in very highly resistant colours (Pfaff, 1973).

There is no denying that fast firing of up to 1280°C has resulted in a widening of the available colour palette for those temperatures; however, they are too high for certain colours, such as the cadmium/selenium reds. For this reason a new type of colour stain has been developed for medium temperature in-glaze fast firing of hard porcelain, i.e. 900–1000°C, extending to practically the whole range of on-glaze colours. These new stains do not only cover an extended palette but are also more brilliant. Despite the lower temperature, they still sink into the glaze and are claimed to be equal to the high temperature fast fire in-glaze stains in respect of chemical durability. The glaze is still highly viscous at the maximum temperature of the medium fast fire stains, so that no air bubbles are formed, the glaze surface remaining perfect; with high temperature fast fire in-glaze colours there is a danger of pin-holes being formed in the glaze (Pfaff, 1985).

Special burnished gold preparations have been developed for firing with the medium temperature fast fire stains. The gold does not actually sink into the glaze, otherwise it could not be burnished and would remain dull. (Golds have, nevertheless, been prepared for high-temperature fast firing (Anon., 1975—German patent 2 208 915).)

The question has arisen as to how the body would stand up to the drastic thermal shock in the fast heating and cooling of fast fire for decoration. Thick-walled hard porcelain articles tended to show cracks after exceedingly fast speeds of firing; prolonging the firing time eliminated this fault. When investigating any possible changes in body properties arising from fast firing, it was found that a certain improvement in the mechanical strength of hard porcelain was achieved; this could be the result of a tempering action as observed with tempered glass.

Regarding pottery, such as bone china, having a thermal shock resistance inferior to hard porcelain, no insuperable difficulties were experienced with fast firing of decoration. Where cracking did occur, a relatively small adjustment in the firing time overcame the fault.

Further Reading

General: **Shaw (1968)**; Singer and Singer (1963, pp. 797–831).
Atomic basis of colour: Dinsdale (1986, p. 184).
Formation of ceramic colours: Shaw (1966); Taylor (1967); Singer and Singer (1963, pp. 227–235).
Colouring agents: Singer and Singer (1963, pp. 616–643); Wolf (1937).
The chemistry of zirconium silicate pigments: Batchelor (1974).
Coloured glazes: Mettke (1984).
Factors affecting lithography: Scherez (1979).
Mechanized and automated decoration:
    Banding machines; Ellis (1973); Alt (1972)
    Printing machines; **Roberts (1981/4, 1981/5)**; Alt (1972); Luchs (1985); Basnett (1980).
Fast firing of decoration: Bull (1982); Pfaff (1973, 1985); Schüller and others (1977); Anon. (1975); Hauschild (1978).

9

Types of Pottery

There is hardly any branch of industry where the very term describing the industry concerned can have so many different meanings to different people as the pottery industry. Some think of pottery as the ware produced by the peasant potter or the studio potter. According to Dodd (1967):

> the term is generally understood to mean domestic ceramic ware, i.e. tableware, kitchenware, and sanitary ware, but the pottery industry also embraces the manufacture of wall and floor-tiles, electroceramics and chemical stoneware . . .

In the present context pottery includes the following ceramic ware:

Artware (ornamental pieces, fancies, figures, figurines)
Tableware (dinner, tea, and coffee services)
Cooking ware (ovenware, fireproof ware, flameproof ware)
Kitchenware

Tableware comprises by far the largest section. A distinction is made between cups and hollow ware such as teapots, jugs, etc., and flatware, e.g. plates, saucers, and meat dishes, usually referred to as "flat".

The most important types of pottery and their main characteristics have already been shown in Table 1.1 (Chapter 1, p. 3).

Table 9.1 lists the whole range of pottery (not given in order of importance) relative to applications.

Kitchenware, not mentioned in Table 9.1, is made of common pottery, often with an engobe, or of earthenware or stoneware.

In an attempt to clarify pottery a division is usually made between *porous* and *dense* bodies and these are then subdivided into *coloured* and *white* bodies. According to A.S.T.M. (American Society for Testing and Materials) C242, white bodies of fine texture are termed *whiteware*.

The definitions of the various types of pottery are just as ambiguous as the word "pottery" itself, if not more so. Besides, a certain overlapping is unavoidable.

TABLE 9.1. *Application of Types of Pottery*

Type	Artware	Tableware	Cooking ware
Common pottery (coloured earthenware, terra cotta (unglazed))	X	X	X
Majolica (faience)	X	X	
Earthenware (fine earthenware)	X	X	X
Lithia pottery			X
Stoneware	X	X	X
Wedgwood's jasper and black basalt	X		
Semi-vitreous and vitreous china		X	X
Belleek china	X	X	
American fine china		X	
English translucent china		X	
Parian	X		
Hard porcelain	X	X	X
Bone china	X	X	
Cordierite porcelain			X
Glass ceramics[a]		X	X

[a] Not pottery, but the end product is like pottery.

1. Porous Bodies

1.1. *Common Pottery*

Common pottery, often called earthenware by the non-expert, comprises a very wide range of ware. The adjective "common" is misleading because in the sense of being commonly used it cannot touch the popularity of *white* earthenware, and because it is not common in the sense of being vulgar; on the contrary, it includes perhaps the most artistic creations of all pottery. Many great artists, including Picasso, have found it a fascinating medium. It was, indeed, the first medium in which man expressed his artistic feelings. The first potter used clay as he found it and this is still true today in respect of primitive peoples like certain Indian tribes in South America. Even in our own society there are still craft potters who dig up their clay in their back gardens.

Centres of pottery have developed around suitable clay deposits. The "Potteries" in North Staffordshire are a good example. This district has remained the centre of the pottery industry, the common pottery originally made there having been ousted by more "refined" ware. North Staffordshire potters "import" most of their clays from Cornwall, Devon, and Dorset, whereas the studio potters in these clay-rich counties (as well as in others) obtain their clays, the famous Etruria marls, from North Staffordshire.

Common pottery, particularly that made in studio potteries, is thrown or moulded, decorated in the clay state, usually by slip trailing, and fired to about 900°C ready to take the glaze. This is applied by dipping, less often by spraying or brushing. The ware is then finally fired to about 1000–1100°C, body and glaze being matured at the same temperature. Normal earthenware lead glazes are used (glazes 6, 7, and 8, p. 133).

Some artware is neither glazed nor decorated, form being the sole criterion of beauty.

Common pottery is porous and therefore comparatively light in weight. It is mechanically rather weak. This is counteracted by its being rather thick walled.

1.2. Majolica

Majolica, sometimes referred to as faience, is a refinement of common pottery. The body does not usually consist of clay only as with common pottery, but contains sand and possibly fluxes. Manufacture follows the same lines as common pottery. It is always glazed.

The two terms are based on geographical locations where this type of pottery was first made towards the end of the fifteenth century. Faience is derived from the Italian town of Faenza and majolica from the island of Majorca.

The definition of faience has changed its meaning. According to French terminology faience is a collective word for all porous pottery bodies, no matter whether white or coloured, that carry a glaze. The faience which used to be made at Faenza and which reached its golden age in the sixteenth century is characterized by a tin-opacified white glaze.

Majolica, also called common faience, has a porous, coloured body which (like faience according to the older meaning) is also covered with a white opaque tin glaze or possibly with a white engobe and a transparent glaze (Fig. 9.1).

One of the most famous types of pottery, *Delft ware*, known by its characteristic blue decoration (also named after the place from where it stemmed and which is still made exclusively in that famous town near The Hague), is sometimes classed with majolica, since it too has a tin-opacified glaze. However, the Delft body is lighter in colour and can, therefore, be regarded as earthenware. In the seventeenth and eighteenth centuries the Delft potteries tried to imitate Chinese porcelain and it is said that while failing in doing so they created the most beautiful earthenware ever made.

Majolica-type bodies, even if not known by the ancient name, are still being made today by studio potters. Majolica is always artistically satisfying.

1.3. Earthenware

The bulk of ceramic tableware consists of a white or near-white body. Like all whiteware, earthenware (always glazed) is almost exclusively made industrially; it is fabricated on a mass-production scale and has lost the

FIG. 9.1.   Majolica plate, St. George after Donatello (Caffaggiolo), about 1515–20, an example of the high standard of pottery decoration in Renaissance Italy. By courtesy of the Board of Trustees of the Victoria and Albert Museum.

individual stamp of the craft potter who, in face of competition by mass-produced articles, does not usually attempt it.

To the non-potter, "earthenware" vaguely means the cheaper kind of common pottery. The name is associated with pottery that has a non-white body. This seems reasonable because "earthen" ware, coming from the brown earth, would, naturally, not be white. However, according to the accepted definition, earthenware has a white body, or rather near-white, body which is porous and thus permeable to liquids and gases. Very often the non-expert calls earthenware china because of the erroneous assumption that all white pottery is china. The definition of china implies translucency, which is absent in earthenware.

Sometimes, in order to bridge the gap between experts and laymen, what (according to the definition laid down) is regarded as earthenware (white or near-white earthenware) is described as "fine" earthenware and what to the housewife is earthenware as "common" or "coloured" earthenware.

On the continent of Europe earthenware is divided into three groups: clay earthenware, lime earthenware and feldspathic earthenware.

Clay earthenware, also dscribed as argillaceous earthenware, according to Bourry (1919), is the oldest earthenware emerging directly from common pottery. It has been made since the twelfth century. It originally consisted entirely of a clay not too high in iron but fairly rich in other fluxing impurities. Later, flint, sand, or other types of quartz were added.

Lime earthenware (calcareous earthenware—Bourry, 1919) is either made of a natural marl (Chapter 2, p. 19) high in lime, with or without the addition of free silica in one form or another, or of a clay to which lime is added as one of the carbonates e.g. limestone, chalk, or marble. Dolomite is also sometimes used. This type of earthenware is highly porous and therefore light.

Feldspathic earthenware, also referred to as "hard earthenware" contains 5–20 per cent feldspathic materials, not usually feldspar itself, but pegmatite or Cornish stone (Chapter 2, p. 31), the remainder consisting of approximately 50 per cent of clays, both ball clays and china clays, and up to 45 per cent of free silica, usually in the form of calcined flint.

Feldspathic earthenware might be regarded as an English development. Its originator was the North Staffordshire potter Astbury. According to the usual account he was struck by the whiteness of flint used by a veterinary surgeon and therefore tried it in his earthenware body. It made the body both whiter and harder.

The man who really laid the foundation for the British earthenware industry was Josiah Wedgwood; he created a cream-coloured body which was so much admired by Queen Charlotte that it was called "Queen's ware". The important point is that in the earthenware first produced by Wedgwood a portion of the clay was introduced as china clay, the then new material discovered by Cookworthy in the St. Austell district of Cornwall in 1768.

Feldspathic earthenware is fired to a higher temperature (1050–1150°C for biscuit and 900–1050°C for glost—glazes 6, 7, and 8, p. 133) than the two other, much less important, types of earthenware.

Feldspathic earthenware, the mineralogical constitution of which has already been described in Chapter 6 (p. 94), is much stronger than clay and lime earthenware. Very high-quality earthenware is even tougher than some hard porcelain because it is less glassy. In Britain, earthenware occupies almost the same position as the more expensive hard porcelain on the continent of Europe. It is particularly appreciated for its under-glaze decoration. Its basic chemical composition has virtually remained unchanged since the eighteenth century, but traditional British ingredients have, to a large extent, been replaced by materials similar to those used on the continent of Europe, i.e. Cornish stone by feldspars and flint by quartz sand.

After World War 2 difficulties arose in obtaining flint free from excessive inclusions of chalk; moreover, flint (derived from the pebbles on the sea shore) is very hard and has to be calcined to make it friable (see Chapter 2, section 2.1.2. pp. 26 and 40) before it is ground which adds to the cost. Sand is not normally calcined prior to grinding. A straight replacement of flint by sand of identical fineness occasionally led to mismatch with the glaze, as shown by impaired crazing resistance. This problem has been surmounted by an adjustment of the biscuit firing temperature, or by a change in the amount of sand or, most expediently, by altering the fineness of the sand.

Earthenware, being porous, tends to absorb moisture if exposed to it for a long time. The moisture attacks the glass in the body and causes an expansion which, in turn, puts the glaze in tension and can thus cause crazing (Chapter 7, p. 126). Moisture expansion can be avoided by introducing wollastonite in the body (Chapter 2, p. 33). This is thought to promote the formation of crystals which are not attacked by moisture in the way glass is subjected to chemical reaction with water. Penetrating the glass, the water breaks some of the network forming bonds and loosens the structure.

1.4. *Lithia-based Cooking Ware*

Some types of earthenware or common pottery can be used for cooking food in the oven. However, they must never be placed directly on a heat source such as a gas ring or electric hotplate.

In an effort to increase resistance to heat and to thermal shock, bodies have been developed with very low thermal expansions. This is brought about by the presence of crystals of the lithia-based minerals (Chapter 2, p. 32). As far as the author is aware, only one factory (Gustavsberg in Sweden) produces such a type of porous cooking ware in Europe. It is claimed to be *flameproof* and thus suitable for cooking on a direct heat source.

It is not difficult to produce *porous* bodies of very low, even zero or negative expansion by mixtures of clay and petalite. However, within the framework of normal pottery practice, it is virtually impossible to obtain a glaze to match such a low-expansion body. In order to reduce thermal mismatch the glaze coat has to be extremely thin, glaze being sprayed on the unfired pieces and matured in one firing together with the body. The firing temperature is very critical, slight increases cause the lithia-based mineral crystals to be taken up in the glass and this causes a sharp rise in thermal expansion.

As a result of a special heat treatment it was possible to use compositions in the systems $Li_2O/ZnO/Al_2O_3/SiO_2$ as "glazes" for bodies of thermal expansion coefficients lower than 2×10^{-6} (20–1000°), as has already been referred to in Chapter 7, end of section 3.1, "The Significance of Thermal Expansion", p. 126. The special heat treatment follows the principle of glass ceramics (see below under section 2.4, p. 193).

2. Dense Bodies

2.1. *Stoneware*

Stoneware evolved in a manner similar to common pottery. The big difference is that stoneware is dense, vitrified. Like common pottery, stoneware arose in places where there were deposits of clays specially suitable for potting. Stoneware, perhaps more than any other type of pottery, is associated with localities. The particular characteristics of local stoneware arose from the clay deposit alone, not the technique used, as in Faenza, Majorca, and Delft. Stoneware clays are unusually rich in fluxes and provided that the temperature of firing is high enough they cause the body to become vitrified and thus non-porous. The same effect can be obtained by adding fluxes to the clay. The stoneware body is always coloured, sometimes brown, often light grey. Stoneware, unlike common pottery, faience, or majolica, is occasionally fired reducingly; this alters its colour.

Firing temperature varies from below 1100°C to about 1300°C, depending on the flux content. The glaze can be applied after the high-temperature firing of the body and matured at a lower temperature (1000–1100°C lead glaze, e.g. glaze 6, p. 133) or to the low biscuited body and matured together with the body (leadless glazes 3, 4, and 5, p. 132).

Stoneware is the only type of pottery that sometimes carries a salt glaze. Early eighteenth-century English salt-glazed stoneware figures are fine examples of this type of glaze. Salt glazes (described in greater detail in Chapter 7, p. 136) are brought about by vapours produced towards the end of the firing process. Salt glazes, being thin, are less smooth and do not always hide the imperfections of the body as do other thicker glazes.

Stoneware has a harsh, almost a rough beauty that the non-expert often fails to appreciate. He does not realize that it is bound to be dearer on account of the higher firing temperature and tends to regard it as unjustifiably expensive. Furthermore he finds it too heavy, which, of course, is due to the lack of porosity. For the same reason it has a very much higher mechanical strength than common pottery. Stoneware has gained in popularity since the nineteen-seventies, mainly because of its "rustic" appeal. The buying public had become aware of the beauty of studio pottery stoneware, generally much more artistic than mass produced industrially made articles. Industrial potters, aware of the new trend, started to introduce stoneware which looked like studio pottery but could be manufactured much more cheaply under mass production conditions. (Moreover, as indicated in Chapter 6, section 7, "Once Firing", p. 122, stoneware is particularly suited to once firing.) The connoisseur will still be able to differentiate between individually made and mass produced ware.

In the past, stoneware has undergone special refinements and in this no one was more successful than Josiah Wedgwood. His lasting achievements were *jasper* and *black basalt*. They are still being made today.

Jasper, although usually regarded as a special refined type of stoneware, could be described as a porcelain, because unless being coloured by one of the usual colouring oxides, it is practically white and translucent when thin.

Fifty per cent of the body consists of the usual pottery ingredients—clay, Cornish stone, and flint; the remaining 50 per cent are heavy spar (barytes). Jasper is the only ceramic body which contains a sulphate as a major constituent. Sulphates in pottery, for reasons not fully understood, cause faults (particularly affecting the glaze) during firing.

Jasper is associated with a characteristic type of decoration: relief figures in white body are stuck on to the main piece which is made of a coloured body, notably of the beautiful shade of blue which derives its name from Wedgwood. It is *pâte-sur-pâte* decoration; the delicate white relief figures are translucent and to this jasper owes its particular charm (Fig. 9.2).

Jasper, mainly used as ornamental ware, is sometimes white (and glazed) inside, only the outside being coloured; the figures are always white. The latter are made separately in hard earthenware moulds. The body is pressed into the negative relief, smoothed over, and the pressed figures taken off the relief in the mould with a type of spatula. Their removal from the mould and affixing to the main piece requires great skill.

The relief figures are noted for the wonderful detail. This is found nowhere else in pottery and is due to the great hardness of the high-fired earthenware moulds used, which, unlike plaster moulds, do not wear. They are made in the first place from a large model on which the fine detail is worked out. The mould made from this large model is then fired, which causes it to shrink. Another model is made from this shrunken mould and this smaller model in turn used for making yet another mould which can again be fired. Thus smaller and smaller editions can be made, thereby retaining the exceptional sharpness of detail throughout. The ceramic moulds are refractory so that there is no blurring of the outline of the design during firing. It is interesting to note that ceramic moulds were used by the ancient Romans.

Black basalt, made of a jet-black body, is almost invariably left unglazed outside and only sometimes glazed inside. It consists of ball clay or red clay and high amounts of manganese oxide and/or calcined ochre, both of which not only act as powerful colouring agents but also as fluxes.

2.2. *Porcelain and China*

This large group of pottery represents the potter's greatest achievement: a dense body that is not only white but translucent. Here confusion over definition is even worse than with the terms "pottery" or "earthenware". First, what is the difference between porcelain and china, if any?

According to technological definitions (Dodd, 1967), in Britain "porcelain" signifies hard porcelain whereas "china" is synonymous with bone china.

FIG. 9.2.   Vase of green and white jasper, Wedgwood, Etruria, about 1785, showing exquisite detail of relief modelling. By courtesy of the Board of Trustees of the Victoria and Albert Museum.

In the United States the terms "porcelain" and "china" are defined on the basis of use: porcelain is a glazed or unglazed vitreous ceramic whiteware used for technical purposes, e.g. electrical porcelain, chemical porcelain, etc. (Dodd, 1967),* according to A.S.T.M. C242, china means any glazed or unglazed vitreous ceramic whiteware used for non-technical purposes, e.g. dinnerware, sanitary ware, and artware, provided that it is vitreous.

The interpretation of the British pottery technologists is not shared by the general public in this country. The British housewife regards porcelain as something vaguely superior to china. The word "china" means to her any

---

* In *Ceramic Glossary* (Schoick, 1963) it is stated: "In countries other than the United States, it is used as a synonym for china."

pottery that is white, which, of course, includes earthenware, although she certainly appreciates the difference between ordinary china and bone china.

The term "china" is derived from "chinaware", which was a collective expression for pottery ware imported from China (or other oriental countries such as Persia); most of it was white and translucent. The affix "ware" was dropped and anything which looked vaguely like the Chinese ware was called "china". The connoisseurs abhor the word "china" and even speak of "bone porcelain".

The word "porcelain" has its origin in the Italian "porcella" literally "little pig", a Mediterranean sea-snail, whose shell is white and translucent. Marco Polo was the first to apply the name to porcelain. Solon (1907–8), an eminent potter and authority, once remarked:

> In the production of a porcelain body a curious similarity of external appearance has been obtained by so many different combinations of constitutive materials that the term does not admit of a scientific definition that would equally suit the numerous varieties of the ware.

Porcelain is thus a collective term comprising all ceramic ware that is white and translucent, no matter what ingredients are used to make it or to what use it is put. Bone china is just as much a type of porcelain as is hard porcelain since both are white and translucent ceramics.

European continental ceramists, more often than not, translate *porcelaine*, *porzellan*, *porcellana*, etc., as "china". The term "china" should be restricted to mean "bone china" as defined by Dodd (1967). However, the fact has to be accepted that it is still widely used to describe a number of other types of porcelain.

2.2.1. Semi-vitreous China and Vitreous China

Vitreous china (also called vitrified china, in Britain usually referred to as vitrified earthenware or vitreous hotelware, if used as tableware) and particularly semi-vitreous china, show translucency, the main characteristic of porcelain, only if thin walled.

In the approach to semi-vitreous and vitreous china the frontiers of the various types of pottery become blurred. Definition has almost ceased to be meaningful and the names given to this type of ware depend more on geographical locations, use of language, and historical factors and associations than technical criteria. The designation "vitreous or vitrified china" stems from the popular notion of all white pottery being china and is a misnomer, since, according to the official United States as well as British definitions, china implies that the body is vitreous in any case.

If in earthenware the content of feldspathic flux is increased or if the biscuit firing temperature is increased the porosity is reduced and the product becomes denser and stronger. This was done in England, where the resulting ware became known as "white granite" or "ironstone china",* also "flintware" (in Scandinavia it is still known under that name). In the United States of America a similar development resulted in "semi-vitreous" and eventually "vitreous" china.

Ironstone china was very near to fine stoneware in character, colour being the only difference which, however, was merely marginal.

The step from earthenware to semi-vitreous china and hence to vitreous china is a perfectly logical evolution, semi-vitreous china having a low porosity (water absorption 2–3 per cent according to Camm and Walters (1983), corresponding to a porosity of approx. 5–7 per cent), vitreous china being almost free from open or interconnected pores. As mentioned, these transitions are brought about by increases in firing temperature or greater flux content or both.

In the United States, vitreous and semi-vitreous china, besides earthenware, are the staple type of pottery (or rather "whiteware" as it is referred to there) produced. The ingredients are basically those of earthenware (on average 47½ per cent clay, 37½ per cent quartz, 15 per cent feldspar), although feldspar has been largely replaced by nepheline syenite, which is available in great abundance in the States and Canada; its greater fluxing power allows a reduction in firing temperature which in turn decreases the cost of the product.

Steatite or wollastonite are occasionally introduced in small amounts, providing a very powerful fluxing action in conjunction with the feldspathic materials. In rare instances lithia-bearing minerals are used.

In Britain, as in the case of earthenware, Cornish stone has been replaced by feldspar itself or other feldspathic minerals, and flint by quartz sand. The sand body was found to vitrify at a lower temperature than a body with an equal amount of flint of identical grain size distribution. It was postulated that this was not due to the greater reactivity of the sand but to its giving a less porous unfired body than flint (Johnson, 1976).

As with most types of pottery, the body of semi-vitreous and vitreous china is first matured at a higher temperature (1100–1200°C). The glaze is of the low-temperature type, similar to or harder than that for earthenware, maturing at about 950–1100°C (glazes 6 and 7, p. 133). Semi-vitreous and, still more, vitreous china is mechanically strong, surpassing normal hard porcelain and earthenware. The glaze surface is better than on earthenware because of the density of the body, and also more brilliant than on most hard porcelains. The microstructure of vitreous china is nearer to that of earthenware than to that of (well fired) hard porcelain.

---

* The ironstone china made by C. J. Mason & Co. (1829–45) contained blast furnace slag (Anonymous, 1966) a very "new" ceramic material now being exploited using glass-ceramic techniques.

2.2.2. The Soft Porcelains

Soft porcelains, or *soft paste porcelains*, are largely of historical interest only and are brought about by a number of different means which, however, produce similar effects.

A characteristic of most soft porcelains is their pronounced brittleness. However, the reason for being called "soft" is their lower firing temperature compared with hard porcelain. (Bone china is relegated to the soft porcelains in some textbooks since its firing temperature is in the soft porcelain range.)

Lower ("softer") firing temperature implies higher flux content in the body. This does not only mean greater brittleness but also less clay, which in turn reduces plasticity and therefore makes production difficult. Furthermore, clays free from impurities, e.g. china clays, have to be used to guarantee whiteness and translucency. These are usually less plastic than more contaminated ones.

It has been shown that porcelain generally was not an invention but the outcome of an evolutionary process in the potter's striving for perfection (Rado, 1964). Porcelain was first made in China as early as the tenth or ninth century B.C. and it was most probably a type of soft porcelain. Different types of porcelain, including hard porcelain, were made at different times and different places in China. It was exported and Europeans first came in contact with it during the Crusades.

Frit porcelains.[*] Europeans endeavoured to imitate Chinese porcelain and the first successful attempts took place in Italy in the sixteenth century. Francesco de Medici of Florence was the first to make porcelain in the Western world. It is known as "Medici porcelain" (Fig. 9.3). Exact records of the composition of the body and the glaze and their preparation have been preserved.

Sand and "saltpetre" were fritted (melted) and mixed with "marine" salt, soda ash, plaster-of-Paris, and alum. This mixture was again fritted and subsequently mixed with clay and plaster. The final body was analysed by percentage as follows:

Silica	73.08
Alumina	10.64
Ferric oxide	1.36
Lime	3.10
Magnesia	0.43
Potash, soda (by difference)	11.39
	100.00

The high silica and alkali contents and low percentage of alumina are striking features of this body. It is like a glaze maturing at about 1300°C

[*] Some authorities regard frit porcelain as the only true soft paste porcelain.

FIG. 9.3. Pilgrim-bottle, Medici frit porcelain, Florence, about 1580, the first porcelain made in Europe. By courtesy of the Board of Trustees of the Victoria and Albert Museum.

(falling between glazes 2 and 3, Chapter 7, p. 132), as can be seen from the following calculated molecular formula:

$$\left.\begin{array}{l} 0.23\ CaO \\ 0.05\ MgO \\ 0.72\ K_2O,\ Na_2O \end{array}\right\} \left.\begin{array}{l} 0.51\ Al_2O_3 \\ 0.04\ Fe_2O_3 \end{array}\right\}\ 5.97\ SiO_2.$$

Medici porcelain is a remarkable technological achievement. It is a combination of glass and pottery and can be described as "milk glass" of porcellaneous appearance, fabricated by ceramic methods.

After Francesco's death the art of making frit porcelain in Europe seems to have been lost and no more porcelain of this type was made in the provinces of Italy. Almost 100 years later (1673) Louis XIV granted a patent to Louis

Poterat to make porcelain in Rouen, which was similar to Medici porcelain. The making of frit porcelain spread to other cities in France, e.g. Saint Cloud, Paris, Sèvres, and eventually to England where potters in Bow, Chelsea, and Derby (all places outside the "Potteries" area in North Staffordshire) became its foremost exponents. John Dwight who, starting from stoneware, had already made "salt-glazed porcelain", succeeded in about 1670 in developing a frit porcelain. In 1671, 2 years before Poterat's "discovery" in France, he applied for a patent which cites that he discovered the "Mystery of Transparent Earthenware commonly known by the names of Porcelaine or China and Persian ware . . .".

With the discovery of china clay in the early eighteenth century in Saxony and the 1760s in France and England, frit porcelain was gradually ousted in Britian because of the difficulty in making (through low content of clay), firing (it becomes very soft at the maximum temperatures), and because of its brittleness.

The only type of frit porcelain made in Europe today is *Belleek china* which is produced at the little town of that name in *County Fermanagh*, Northern Ireland. It is also made in the United States, sharing with *American fine china* a composition with a relatively large proportion of glass frit in addition to small quantities of clay, quartz, and possibly whiting, resulting in very high translucency.

Magnesia porcelains. The eighteenth century was a golden age of experimentation in pottery, especially in England, and in the effort to reproduce oriental porcelain many materials were tried. One of the most successful was "soapstone", now known as talc or steatite (Chapter 2, p. 33), which was first used at Bristol in the early eighteenth century and later at Worcester (Fig. 9.4). (In Worcester several different bodies were probably used simultaneously: frit porcelain, bodies containing small amounts of bone, besides the magnesia-based porcelain containing up to 40 per cent soapstone.) Talc was used also by the porcelain factory in Zurich soon after Worcester (around 1764).

Another magnesium silicate, *sepiolite* $(3MgO.4SiO_2.5H_2O)$, found in Spain, was used in the porcelain works of Madrid and Sèvres.

High feldspar porcelains. Bodies consisting of the three "classical" ingredients make up a large group of the soft porcelains, in the approximate proportions of 30–40 per cent clay substance, 25–35 per cent quartz, and 30–40 per cent feldspar. Some ceramists apply the term "soft porcelain" to this group alone. It includes *Seger porcelain*, developed by Seger during the last century, and *Japanese porcelain*. Seger porcelain and certain other soft porcelains are fired in the manner of hard porcelain, viz. biscuit at 850–1000°C and glost at 1250–1300°C (glaze 3, p. 132).

Singer and Singer (1963) also class *American household china* and *English*

FIG. 9.4.   Sauceboat of soapstone porcelain, early Worcester. By courtesy of the Governors of the Dyson-Perrins Museum Trust.

translucent china as soft porcelains. These are really developments of vitreous china, being made fully translucent by higher firing temperature and increased flux content. In American household china the quartz (22 per cent) and feldspar (36 per cent) contents have virtually been reversed compared with vitreous china. As with vitreous china, feldspar is now largely replaced by nepheline syenite, often with the addition of a small amount of talc to increase the fluxing action. (The composition of English translucent china, which was introduced only in 1959, has not been published.)

An unusually high amount of feldspar, about 70 per cent, is used in *parian*. Unlike all other porcelains it is unglazed. The body itself is self-glazing and has a satin-like finish. In order to enhance its fluxing power, the feldspar is calcined (fritted). Parian contains about 30 per cent china clay or less, and normally a small percentage of *cullet*, scraps of glass from glass works. It is used almost exclusively for making ornamental ware which is slip cast so that the relatively low amount of clay is not detrimental. The firing temperature is about 1200–1250°C. Parian was first brought out in the nineteenth century and is seldom made today.

Lastly, *semi-porcelain* might be included in the group of high-feldspar soft porcelains. According to some experts' definition it is slightly porous, having been fired to a low temperature, but its feldspar content is high enough to make it just translucent.

2.2.3. Hard Porcelain

Hard porcelain or *hard-paste porcelain* is sometimes referred to as the only *true* porcelain. (Occasionally it is called "feldspathic china", but this is unsatisfactory because vitreous china is also feldspathic.) It is the most important type of pottery manufactured on the continent of Europe. When anyone on the continent of Europe or the pottery technologist in Britain speaks of porcelain he means *hard* porcelain.

The first hard porcelain was made in China. As far as is known, china* was never made in China. Marco Polo left a vivid description of the exquisite ware "made from the crumbling earth and clay that was left for 30 or 40 years exposed to rain, wind, and sun". The first detailed account of the art of making porcelain in China was provided by the letters in 1712 and 1722 of the Jesuit missionary, Père d'Entrecolles. The Chinese used kaolin and *pe-tun-se*, a type of pegmatite, for glaze as well as body (in different proportions, of course). The glaze must thus have been a high-melting one as is used in true hard porcelain, or at least Seger porcelain.

The originator of hard porcelain in Europe was Böttger, an alchemist in the service of the Elector of Saxony, August the Strong, at Dresden. (Hard porcelain is often referred to as "Dresden china" in the United States.) It is significant that an alchemist, not a potter, was entrusted with the task of producing porcelain. After working virtually as a prisoner for nearly 10 years he succeeded in this in 1709, thus without the benefit of Père d'Entrecolles's letters. The porcelain he made was not of the hard porcelain composition we know today. He did not use feldspar but calcareous fluxes, viz. alabaster and chalk. Feldspar was introduced into Böttger's body in 1720. The significant point was his realization that the porcelain imported from China at the time was not an opacified glass (the basis of the frit porcelains) but was made according to pottery methods using a mixture of special earths. As a result of a survey of raw materials, particularly clays, in the Province of Saxony, arranged by his princely employer, he stumbled on china clay (kaolin), the material essential for making porcelain.

Mields (1965) has drawn attention to the fact that the same materials as used in Böttger's porcelain were found in earthenware also produced by Böttger. He may have noticed that near the entry of the flame into the kiln where the temperature was highest there was a vitrified piece which gave him the clue that by greatly increasing the firing temperature he could produce porcelain. (It is known that the temperature in eighteenth-century kilns was so uneven that body and glaze composition had to be varied to suit particular positions in the kiln.)

Again, more or less by accident, he may have arrived at his glaze which, like that of the Chinese, was a high-temperature one maturing together with the body contrary to all pottery glazes made in Europe up to then which were for low temperatures. The composition of the glaze suggests that it was a result of a body experiment containing unusually high amounts of quartz and chalk.

The discovery of hard porcelain was independently repeated by Greiner in

---

* Bone china.

Thuringia and Vinogradoff in Russia. Around 1768 Cookworthy made the first hard porcelain in Britain following his discovery of china clay in Cornwall.

The manufacture of hard porcelain was taken up in many countries in Europe during the eighteenth century. In England it was not made on a large scale and its production ceased altogether in favour of bone china in the early nineteenth century. The fact that the making of hard porcelain in Britain never reached the significance it did on the continent of Europe may be ascribed to various reasons. The most important one was probably the lack of suitable refractory clays for saggars. Most of the British fire clays are rich in iron, which does not matter so much (in fact may even help) for normal pottery fired in an oxidizing atmosphere. However, because of reducing conditions in the firing of porcelain, low-melting ferrous silicates are formed which can cause the saggars to collapse during firing. Another reason may have been lack of suitable coal. British coal, being very pure, did not lend itself to producing the long flames required for reducing firing as did the gaseous, continental *brown coal*. A further factor was possibly lack of suitable materials apart from china clay. It is highly likely that Cookworthy used Cornish stone since it is similar to *pe-tun-se*, the Chinese pegmatite. Cookworthy's porcelain is reported to have been "marred by smoke-staining" caused during firing. This was probablay due to the fluorine present in Cornish stone. Such greyish discolorations have been more recently observed on porcelain made with Cornish stone. Other suitable feldspathic minerals are not available in the British Isles. Finally, in view of bone china and the high-quality earthenware produced there was no real need for hard porcelain. Today the only hard porcelain in the British Isles is made in Worcester (Fig. 9.5). Outside Europe hard porcelain is produced in large quantities in Japan. It is not manufactured in America, although Gould (1947), in his book on hard porcelain, made a strong case for its manufacture being taken up in the States.

The classical composition of hard porcelain is: 50 per cent clay substance, 25 per cent feldspar, 25 per cent quartz.

Wide variations in the composition are possible, e.g. bodies for cooking ware are higher in clay substance and lower in quartz. Highly translucent hard porcelains are richer in feldspar and quartz. There have been no changes in the basic chemical composition of hard porcelain. However, as indicated in Chapter 2, sections 2.2.2, "Feldspar-bearing Rocks", and 2.2.4, "Winning and Beneficiation of Feldspathic Minerals", pure quartz and feldspars from individual mines, now exhausted, have been replaced by quartz and feldspar, produced by froth flotation of pegmatite.

By using a sericite clay (see Chapter 2, section 1.1.2, "The Action of Heat on Clay Minerals", p. 12) in a porcelain body, replacing all crystalline silica, the firing temperature could be reduced and the mechanical properties of the porcelain improved (de los Monteras and others, 1978).

FIG. 9.5.   Royal Worcester pattern "Evesham" tableware and oven-to-tableware in
hard porcelain, present time. By courtesy of Royal Worcester Spode Ltd.

Unlike most other pottery bodies, hard porcelain is first subjected to a low-
temperature firing (900–1000°C), merely a hardening, which makes it strong
enough to withstand being dipped in glaze slip. This "hardening" of the body
arose from the use of the old bottle ovens. The maturing firing of hard
porcelain was (and still is) done in a reducing atmosphere. This implied long
flames and it was a clever move on the part of the old potters to utilize this
potential source of heat (which would otherwise have been wasted) by
building an upper "deck" above the main firing chamber for the biscuiting of

the body. It so happened that the temperature was just right for giving the body sufficient strength, but with sufficient porosity (over 30 per cent apparent porosity) to make dipping in glaze easy. These ovens were known as "double decker" kilns. There were even "treble decker" ovens, having a further storey utilized for firing saggars.

When the porcelain makers of today changed over from bottle ovens to modern kilns they retained the principle of firing the body to a low temperature first. However, as there was no waste heat available, separate low-temperature kilns had to be built for hardening.

Body and glaze are fired to the same temperature. The glaze is not fritted (as it contains no lead or boron compounds) and represents the highest temperature glaze used in all pottery (glaze 2, Chapter 7, p. 132). The firing temperature of hard porcelain is higher than for any other type of pottery (1400°C).

The question arises: Why should an oxidic material be subjected to reduction in its heat treatment? The answer is twofold: (a) to avoid blisters, and (b) to obtain a white product.

In the traditional coal-fired bottle ovens the atmosphere varied from reducing, when coal was placed in the fire boxes, via neutral, to oxidizing, as the coal burned down causing greater and greater access of air until the whole sequence was repeated. The iron oxide in the body was thus subjected to a continuous change from the ferric to the ferrous state. Bearing in mind that enormous amounts of gas were evolved in these changes, blisters were caused after the body had undergone vitrification when these gases could not escape any more.

Alternating oxidizing and reducing conditions invariably cause blisters. A uniform atmosphere throughout the firing does not.

When the first porcelain makers fired their ware normally they probably found that it blistered and was yellowish in colour. If, after they had reached a certain stage in the firing, a temperature slightly in excess of 1000°C, they kept the fire boxes full of coal, they must have discovered that the product fired under these conditions was not only free from blisters but also whiter. This manner of stoking caused a continual deficiency of oxygen, at the same time ensuring an excess of carbon monoxide. This in turn was responsible for the very small amounts of finely distributed iron oxide, invariably introduced by the raw materials, to be converted into ferrous oxide (Berg, 1963) in the final body. As already pointed out (Chapter 6, p. 100), this is dark green in colour but compared with the brown ferric oxide has only a very slight staining effect and gives the hard porcelain its characteristic bluish-white tint.

In vitreous china and most of the soft porcelains the atmosphere is kept oxidizing throughout so that there is no change in the state of the iron oxide: hence no blisters occur but, as the iron is present as ferric oxide with its higher staining power, the ware is more or less cream coloured.

There would be no blistering in hard porcelain fired in up-to-date kilns if the atmosphere were kept oxidizing throughout. But it would be cream coloured or greyish yellow; the discoloration would be more intense than in vitreous china. Perpetuating the tradition, porcelain is still fired reducingly during the critical stages even in modern kilns. This is easy with gas-fired intermittent kilns but poses problems with tunnel kilns. Apart from burner settings also draught and pressure conditions in the kiln have to be closely controlled. These in turn are governed by the kiln fill, the setting on the trucks, the speed of propelling, and even climatic conditions. Blisters can still be a problem in modern tunnel kilns if intermittent oxidations are allowed to occur, say by an insufficiently dense setting of ware, allowing gases to take a wrong course.

The great majority of hard porcelain kilns are fired by gaseous fuels such as town gas, propane, butane, oil, etc. In few instances, where electricity is used as a source of heat, reducing conditions are brought about by injection of a reducing gas. Formerly moth balls were introduced into the reduction section of the kiln.

Details about the mineralogical constitution of hard porcelain have already been given in Chapter 6, section 1.2.1, "The Classical Triaxial System" (p. 94).

2.2.4. Bone China

Bone china is a specifically English product (Figs. 9.6 and 9.7). Until the late nineteenth century when it was introduced in one factory in Sweden it was not produced outside England. It is only since the nineteen-sixties that it has been made, independently, in single factories in America, Canada, Ireland, Germany, Russia and Japan.

The particular fascination of bone china lies in the fact that, unlike all other ceramic products, it contains a material (its main constituent) that is not of mineral origin.

According to Solon:

> . . . The refractory power of phosphate of lime obtained from calcined bones had been known so far back in olden times that it is not possible to trace its first application to ceramic art . . .

Shin bones of cows, oxen, and horses were used as opacifying agents in the production of milk glass in the seventeenth century.

However, calcined animal bone in pottery had no practical significance until Thomas Fry's patent appeared in the middle of the eighteenth century. Accordingly the porcelain factory at Bow was the first to incorporate it in a porcelain body. Chelsea, Derby, and Worcester soon followed. Initially the amount of bone ash added was small and acted as a flux.

The credit of having established the composition of bone china of 50 per cent bone, 25 per cent Cornish stone, and 25 per cent china clay, still used at

FIG. 9.6.   Royal Worcester pattern "Royal Diamond" tableware in bone china,
present time. By courtesy of Royal Worcester Spode Ltd.

present, is usually attributed to Josiah Spode II (1789). By 1794, Spode bone
china was in full production.

Whether bone acts as a flux or as a refractory depends on the proportion of
bone introduced and on the other materials present besides bone. Spode
probably knew that small amounts of bone added to a porcelain body of the
hard or soft porcelain type acted as flux. He may have found that by
increasing the amount of bone he came to a point (approaching a eutectic)
beyond which its effect was reversed, bone then becoming a refractory. Or he
may simply have mixed equal parts of bone and porcelain body, the latter
consisting of 50 parts of china clay and 50 parts of Cornish stone, which in
turn brought in about 30 per cent or more feldspar and 20 per cent or less
quartz.

FIG. 9.7.   Royal Worcester figure "Golden Eagle" in bone china, having a wing span of 1 metre, requiring 97 moulds to make—a superb example of Royal Worcester present-day craftsmanship. By courtesy of Royal Worcester Spode Ltd.

No matter how Spode arrived at the composition of bone china, he produced a body that was commercially successful and was eventually adopted by all porcelain makers in England. However, its production is more difficult than that of other porcelains since its clay content is only 25 per cent instead of 50 per cent or more in hard porcelain and other types of pottery. This low clay content causes the plasticity and green strength to be poor (although as pointed out in Chapter 2, p. 35, bone ash is assumed to contribute a little plasticity). This in turn requires greater skill on the part of the operators. An added difficulty is the narrow firing range. Unlike hard porcelain with its wide firing range (well over 100°C) the bone china body composition lies near a eutectic and for this reason the maximum temperature is very critical, having to be controlled within 20°C. The maintenance of such fine limits in the old bottle ovens seems a miracle to the modern technologist. (If he wanted to know how this miracle was achieved he will find the answer in a paper on the subject by Wilson (1916–17), perhaps the last of the great master potters—one of the rare blends, now practically extinct, of scientist and craftsman—who could fire bottle ovens single-handed.)

The reason for the critical temperature control arises from the rapidity at which the reactions take place in the bone-china body as it approaches its final firing temperature of 1260°C. This causes a sudden pronounced softening within a narrow temperature range. To counteract this and prevent distortion the pieces have to be supported to a greater extent than with any other pottery now made. Plates and other flatware are placed on a bed of powdered alumina; formerly powdered flint was used which, however, caused silicosis, or more correctly pneumoconiosis, and its use has been forbidden by law. More recently, tailor-made setters, designed so that the plate can ride on the setter during shrinkage, have been introduced. Cups and hollow ware are normally supported by rings or plaques of the same body (Chapter 6, p. 101). To prevent sticking the support is coated with alumina.

If the temperature of firing is too high the body becomes distorted and blistered. If the temperature is too low the correct degree of translucency is not achieved, the body having more than 0.5 per cent apparent porosity (highest quality bone china is considered as "under-fired" if it has more than 0.2 per cent apparent porosity). This slight "open" porosity can also impair mechanical strength, glaze fit, etc.

After biscuiting, particles of alumina adhering to the ware are removed by brushing, scouring, or rumbling (Chapter 7, p. 139).

The cleaned ware is dipped in a "low sol" (low lead solubility) or leadless glaze and glost fired to about 1100°C (glaze 6, p. 133). The apparent porosity of the bone china body after the biscuit fire being less than 0.5 per cent, makes dipping in glaze somewhat difficult and requires great skill. Spraying of glaze is applied where possible, e.g. flatware.

Although mechanically much stronger than hard porcelain, bone china—because of its high thermal expansion—cannot be used for cooking ware. Its glaze is less resistant to scratching than the hard porcelain glaze but far more brilliant, allowing a much wider palette of on-glaze colours. It is, as a rule, more translucent and always whiter than hard porcelain.

Details of the mineralogical constitution of bone china have already been given in Chapter 6, section 1.2.2, "The System containing Calcium Phosphate", p. 97.

There have been numerous endeavours at replacing bone ash, particularly in Japan. Natural calcium phosphate rocks did not have the purity required. Chemically prepared calcium phosphates caused insuperable workability problems. Perhaps the most successful attempt was the calcining and grinding of a mixture of dicalcium phosphate dihydrate and calcium carbonate to yield beta-tricalcium phosphate. A ratio of Ca:P between 1.55 and 1.65 was claimed to give a body with properties similar to bone china (Taylor and others, 1979).

2.3. Cordierite Cooking Ware

Amongst the dense bodies, stoneware, occasionally vitreous china, but above all hard porcelain, are used for making cooking vessels which can be placed in hot ovens. The pieces are described as "ovenware", or, in the case of hard porcelain as "fireproof ware".

In recent years a new type of cooking ware based on *cordierite* has been produced which is claimed to be flameproof. Cordierite (Chapter 2, p. 33) has been chosen for its very low thermal expansion (as with lithia-based bodies), bearing in mind that low expansion makes for good thermal shock resistance. Unlike lithia-based cooking ware, cordierite porcelain is non-porous. The firing range of pure cordierite is extremely narrow, viz. within 5–10°C. It has therefore been found expedient to introduce the raw materials talc, clay, and alumina, in proportions which do not give the exact cordierite formula. The resultant bodies do not then consist entirely of cordierite but their firing range is practical for industrial kilns. The firing temperature is around 1380°C. As with lithia-based cooking ware, glazing of cordierite bodies presents a serious problem since glazes of matching (lower than body) expansion are practically non-existent. There is no reason why the expedient of a "glass ceramic" type of heat treatment as applied to lithia bodies (see section 1.4, p. 175) should not also be used to "glaze" cordierite cooking ware. Some cordierite bodies are self-glazing.

2.4. Glass Ceramics

In an effort to reproduce Chinese and other oriental porcelain, glass-makers, especially in France, endeavoured to make opaque glass (milk glass) look like porcelain. This was described as *porcelain de verre*; the pieces were made like glass and should not be confused with the Medici porcelains (p. 181) and the later French *frit porcelain* which were fabricated according to pottery methods.

A few different ways of producing opaque glasses have already been described under "Opaque Glazes" (Chapter 7, p. 134). The famous French scientist, Réaumur (mainly remembered for his temperature scale), in the early eighteenth century subjected pieces of glass which had been packed into a mixture of sand and gypsum to red heat for several days. He thereby caused devitrification in the glass objects, the crystals formed being pseudo-wollastonite ($CaO.SiO_2$), possibly formed by an ion exchange with the gypsum ($CaSO_4.2H_2O$), (devitrification is possible, however, without gypsum). The devitrified objects looked like porcelain and were, in fact, the first glass ceramics produced; they were not described as such, but as "Réaumur porcelain". They were mechanically stronger and had a higher thermal shock resistance than the normal soft porcelains of his day.

It took nearly 250 years before the idea of devitrified glasses, or glass ceramics as they are known today, was taken up again, incidentaly not to reproduce porcelain ware, but primarily with the object of making radomes for missiles.

The first important step in the development of glass ceramics in their present form was the discovery of photosensitive glasses. These contain small amounts of copper, gold, or silver which can be precipitated in the form of very small crystals during heat treatment of the glasses.

Photosensitive glasses can be opacified. When heating a photosensitive opacified glass to a higher temperature than that normally employed in the heat-treatment process, Stookey, during his research at the Corning Glass Works, United States, in the 1950s, found that instead of melting, the glass was converted to an opaque polycrystalline ceramic material.

Glass ceramics are made according to normal glass-making techniques from a normal glass batch to which a nucleating agent, such as a noble metal, titania, phosphorus pentoxide, etc., is added. After annealing, the shaped glass is heated to the *nucleation* temperature to allow the maximum number of nuclei to grow. Nucleation completed, the temperature is raised at a carefully controlled rate. The temperature rise is stopped when crystals form, usually 200–500°C above the softening temperature of the original glass. This *crystallization* temperature is held for the required time and followed by cooling to room temperature.

It may be argued that glass ceramics should not be described as a type of pottery since they are not made by pottery methods. However, the end product has the constitution of fired vitrified pottery, although the proportion of crystals is much higher than in most porcelains or other vitrified pottery and the crystals are much smaller. The great advantage of glass ceramics lies in the possibility of varying the final structure and thus the properties by modifying the composition of the original glass, the nucleation agent, and particularly the heat treatment. For instance, glass ceramics of almost any thermal expansion, ranging from zero to that of a high-expansion metal, can be produced. The very low-expansion glass ceramics are used for cooking ware.

Further Reading

General: Singer and Singer :1963, p. 430); Rosenthal (1949); Wiedemann (1967, 1968, 1969); Rhodes (1973).
Historical:
   Pottery: **Savage (1959)**
   Porcelain: **Savage (1963)**; Solon (1907–8); Rado (1964).
   Chinese porcelain: Ayers (1964); Yule (1921); (Marco) Polo-Latham (1958).
   Medici porcelain: Liverani (1959).
   Eighteenth-century porcleain: Charles (1964); Barrelet (1964).
   Meissen porcelain: Mields (1965).
   Cookworthy porcelain: Tait (1962).
   Origin of bone china: Copeland (1961); Radford (1961); Thomas (1933).
Pottery*: Rhodes (1971, 1973).
Majolica: **Ravaglioli and Vecchi (1981)**.
Fine earthenware: **Camm and Walters (1983)**.
Semi-vitreous china: **Camm and Walters (1983)**.
Vitreous china*: Reh (1966).
Stoneware: **Grahn (1983)**.
Hard porcelain: **Schüller (1979)**; Rado (1971).
Bone china: **Rado (1981)**; Franklin and Forrester (1975); Forrester (1986).

* To be published under Ceramic Monographs 2.1.7 (Pottery) and 2.1.2 (Vitreous china), *Handbook of Ceramics*, Verlag Scmid GmbH, Freiburg i.Br.

10

The Properties of Pottery

Certain characteristics are common to all types of pottery. Broadly speaking, pottery is hard, non-flexible, brittle, not affected by heat or cold, fire or water, or the usual chemicals met with in daily life, not even strong acids. It does not of its own accord deteriorate with time, but is durable and virtually indestructible. This does not mean that objects of pottery cannot be cracked or fractured, or its glaze scratched and worn under constant use. Many people fail to realize that useful pottery, such as tableware, is a consumable commodity. Significantly, the properties of pottery bodies, like those of most ceramics, are determined by *texture* rather than *structure*; this explains some of their apparent anomalies.

Table 10.1 shows the physical characteristics of some pottery. Few numerical data have been published on types of pottery which do not also have some technical uses, e.g. stoneware and porcelain. The data given for the various aspects of strength should be regarded with caution because different sources may have used different testing methods. In view of the overriding influence of shape, figures on chipping resistance have been omitted, except for hard porcelain and bone china, where it is certain that test pieces of identical shape were used. Most of the values are not given in the original units but, for the sake of uniformity and comparability, have been calculated to correspond to SI Units.

(The figures quoted in Table 10.1 for glass ceramics cover a range well beyond that of table and cooking ware. Certain technical ceramics, e.g. alumina, have moduli of rupture higher than those of the best glass ceramics.)

1. Mechanical Strength

1.1. *Basic Considerations of Strength and Brittleness*

The theoretical strength of perfect single crystals and of homogeneous glass can be calculated from the known bond strengths between the constituent atoms. The higher the bond energy the greater the theoretical strength. The measured strengths of crystals of both metals and non-metals are, however, many times smaller than the calculated values, but the reason for the discrepancy is different in the two cases.

TABLE 10.1. *Physical Characteristics of Some Pottery Bodies*

Type of pottery	Earthenware	Stoneware	Vitrified hotelware	Hard porcelain	Bone china	Cordierite porcelain	Lithia pottery	Glass ceramics
Reference	B.C.R.A. (1967)	Singer (1951)	B.C.R.A. (1967)	Salmang (1961)	B.C.R.A. (1967)	Salmang (1961)	Salmang (1961)	McMillan (1964)
Water absorption (%)	6–8	0.2–2.5	0.1[a]	0.0–0.5	0–1	0.0–0.5	0–21	0
Specific gravity	2.6	2.50–2.65	2.6	—	2.75	—	—	2.42–5.76
Bulk density (kg m^{-3})	2200	2030–2480	2600	2300–2500	2700	2100–2200	1600–2300	2720–5760
Compressive strength								
Unglazed (MPa)	—	—	—	392–442	—	275–491	137–412	—
Glazed (MPa)	—	571	—	442–540	—	—	—	—
Tensile strength								
Unglazed (MPa)	—	11	—	23–34	—	25–34	19–37	—
Glazed (MPa)	—	—	—	29–49	—	—	—	—
Modulus of rupture								
Unglazed (MPa)	55–72	34[b]	82–96	39–69	97–111	49–84	29–49	69–343
Glazed (MPa)	—	—	—	59–98	—	—	—	—
Impact strength								
Unglazed (Nm^{-1})	—	1766	—	1766–2158	—	1766–2158	1864–2943	—
Unglazed[c] (Nm^{-1})	—	—	—	1167	2188	—	—	—
Chipping resistance								
Glazed[c] (Nm)	—	—	—	10.9–19.0	54.3–77.3	—	—	—
Modulus of elasticity								
Unglazed (GPa)	55	69	82	69–79	96	69–108	—	79–1370
Linear thermal expansion coefficient:								
20–1000°C ($\alpha \times 10^{-6}$ K^{-1})	—	4.1	—	3.5–4.5	—	1.1–1.5	—	−0.4–+16
20–700°C ($\alpha \times 10^{-6}$ K^{-1})	—	—	—	—	—	—	0.06–0.05	—
20–500°C ($\alpha \times 10^{-6}$ K^{-1})	7.3–8.3	—	7.3–8.3	—	8.4	—	—	—
Thermal conductivity								
20–100°C (Wm^{-1}K^{-1})	1.26	1.57	—	1.16–1.63	1.26	1.98–2.32	—	2.21–5.47

[a] B.S. 4034.   [b] Reference: Kingsbury 1939).   [c] Reference: Dinsdale (1961).

All crystals contain defects in their lattice which are known as *dislocations*. In metal these dislocations can move under relatively low stresses and this movement produces *slip* along certain crystal planes. Ionic crystals and glasses, the constituents of pottery, have much more complex structures than metals, and dislocations in these materials are largely immobile. It is therefore to be expected that in such cases the strengths would approach those calculated from bond energies. However, this is not so, because of the presence of flaws, such as cracks and surface scratches, known as "Griffith's cracks", unavoidable in normally prepared materials. These cracks concentrate the applied stress at the tip of the crack to values exceeding the theoretical strength so that fracture results. In metals such surface defects are not so important because plastic flow can usually take place and blunt the crack tip and thus reduce the stress concentration. In ceramics, on the other hand, these defects (due to lack of ductility) cause an enormous weakening. The practical strength of pottery is only about 1 per cent of the theoretical strength of its components (crystals and glass). Ceramic crystals, because of their high bond energies, have the highest theoretical strength of solid materials, amounting to several million lb/in^2.

Ceramic fibres and whiskers almost reach theoretical strength because they are practically without Griffith's cracks. The cost of incorporating them in pottery would, however, be prohibitive at present.

A further source of weakness in pottery is its heterogeneity, the differing thermal expansions of the phases (mullite and quartz crystals and glass) present causing additional stresses.

1.2. *Assessment of Strength of Pottery Bodies*

Strength can be expressed in various ways. The atomic structure of ceramics and glasses (the immobility of dislocations) implies that they are very strong in compression but weak under tension: their *compressive strength* is well over ten times greater than their *tensile strength*; stoneware is shown fifty times stronger under compression than under tension (Table 10.1). A few years ago it was demonstrated that eight bone china cups were able to support a full-size double-decker bus, one cup having been placed under each of the eight wheels. In the service life of pottery the compressive strength is of no practical significance. To measure tensile strength in pottery is somewhat difficult because it is almost impossible to eliminate stresses other than tensile ones. It has been found that *modulus of rupture* shows values of twice those of tensile strength. The strength of pottery is generally expressed as modulus of rupture which is easy to determine. Since, particularly when three-point loading is employed, stress is concentrated in a limited extent of surface skin, modulus of rupture denotes surface strength rather than volume strength.

A comparatively new concept of expressing strength in ceramics has been adopted by Clarke, viz. *work of fracture*. If shock, thermal as well as mechanical, is applied, the material is caused to store, momentarily, elastic strain energy. Where the fracture stress of the material is exceeded the extent of the cracking is limited by the amount of strain energy stored. Certain factors, such as the size of the crystals in the body, may influence modulus of rupture and work of fracture in opposite ways: modulus of rupture is increased by *decreasing* grain size (Dinsdale and Wilkinson, 1966) whereas work of fracture is increased with *increasing* grain size (Clarke *et al.*, 1966).

In use, failure of pottery tableware mostly occurs as a result of some kind of impact, and mechanical strength is often expressed as "impact" or "chipping resistance". Certain factors, i.e. crystal size, affect edge-chipping resistance and bottom impact resistance of plates in opposite ways.

1.3. *Factors Influencing Strength*

1.3.1. *Structure and Texture*

The mechanical strength of pottery bodies is controlled predominantly by: porosity, crystal/glass ratio, crystal size, type of crystal.

These factors depend on the state at which the reactions occurring during sintering are arrested. This is in turn governed by the composition, impurity content, and particle size of the raw materials, as well as by the temperature and other aspects of firing.

High porosity and high glass content reduce the strength. Thus both the very porous types of pottery (e.g. common pottery and majolica) and the highly glassy soft porcelains are weak. The lack of inherent strength in the former is compensated by greater thickness of the pieces. Ideally there should be no porosity if highest strength is to be obtained. However, there is always some residual (true) porosity (which cannot be eliminated), even in vitrified bodies. This severely restricts the attainable strength. Glass is weaker than most crystals, partly because of the lower atomic bond strength (as evidenced by the low Young's modulus), and partly because of the presence of larger Griffith's defects.

The crystal/glass ratio should be as high as possible for maximum strength. The ideal would be to eliminate glass altogether but this is not possible with conventionally made pottery.

The crystal glass ration should be as high as possible for maximum strength. the original particle size of the filler introduced. However, the strength cannot be increased by decreasing the size of the filler beyond a minimum value (Dinsdale and Wilkinson, 1966).

The crystals should have a regular and low thermal expansion, compatible with that of the glass and should not be subject to inversions as, for instance, free silica.

1.3.2. *Design*

Because it is brittle, pottery is suceptible to failure caused by local stress concentrations due to abrupt changes in geometry (design) or loading. Attainable values in material strength may be completely overshadowed by differences in shape. For instance the chipping resistance can be altered by a factor of 30 to 1 (Fig. 10.1). This is much more than can be done by changes in body composition. Plates should have a semicircular edge for best chipping resistance; sharp edges should be avoided. The direction of the blow has also an important bearing. A plate (A) with an edge that offers a flat towards the pendulum at the point of impact has a much higher chipping strength than a plate (B) with a normal rounded edge if the blow is delivered horizontally. However, with blows from other directions (B) is the stronger plate (Fig. 10.2).

Another example of the great influence of testing conditions on strength is shown by the impact test of A.S.T.M. (American Society for Testing and Materials). Test plates are hit at the back in the centre. If the plate is made thicker in the centre much higher impact strength values are obtained than with a plate of normal thickness in the centre. This is misleading because the plate with the thicker centre is not stronger if the blow is directed anywhere except the centre. It is likely to fail in the same manner as the plate with the centre of normal thickness, since impact (chipping) in use almost invariably occurs at the edge.

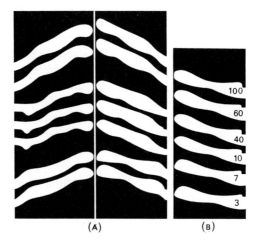

FIG. 10.1. (A) Various edges of plates found in service. (B) Effect on chipping resistance of various edges on one shape of plate. (After Dinsdale, Camm and Wilkinson (1967), p. 384.) Reproduced by permission of The Institute of Ceramics.

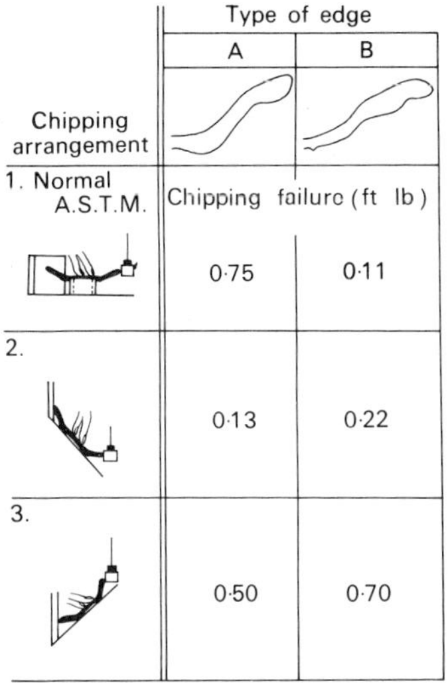

FIG. 10.2.   Chipping resistance of two industrially produced plates. (After Dinsdale, Camm and Wilkinson (1967), p. 389.) Reproduced by permission of The Institute of Ceramics.

1.3.3. *Shaping Method*

Considering "traditional" pottery generally, its texture and hence its strength is greatly influenced by the method of making. Test pieces prepared by different methods, i.e. slip casting, hand pressing, extrusion, etc., give different strength values; higher pressures do not necessarily produce better strength. There is no universally applicable rule as to the superiority of any particular forming method. However, it is reasonable to assume that pottery made by throwing, implying equal pressures, homogeneity and concentric spiralled rings, is stronger than that made by other methods.

Lath-like particles introduced by wollastonite in earthenware bodies in the form of extruded rods or slip-cast discs increased impact and chipping resistance by 300 per cent. However, when incorporated in jiggered (plastic-made) plates there was no increase in chipping resistance. The difference in these results is explained by the orientation of the fibrous laths.

1.3.4. *Glaze*

The body-glaze fit greatly affects the mechanical and thermal strength. The composition of the glaze after firing on to the body is never the same as before firing. This is due to two main reasons: volatilization (only very slight in high-temperature glazes free from lead and boron) and even more so, to interaction with the body, apart from ion exchange between body and glaze. Changes in composition lead to altered expansions which affect the stress relationship between body and glaze and hence their match. It is not usual that interaction alters the glaze to such an extent that its expansion becomes greater than that of the body. However, the reaction layers formed in the interface between body and glaze may have undesirable expansion character-istics which, in turn, could lead to a reduction in strength.

Thus the bone china body is often weakened by glazing but in spite of this the strength of bone china remains remarkably high. The strength of the earthenware body, on the other hand, is usually improved by the glaze. Owing to the porosity of the earthenware body it is easier for the glaze to attack the body. As a result greater reaction occurs than with a dense body and the interface formed is rich in mullite crystals which themselves tend to add to strength.

The surface of the body usually contains flaws, such as scratches or minute cracks which act as stress raisers. The glaze covers up these imperfections. Investigations by Dinsdale *et al.* (1967) showed that in broken unglazed rods nearly all fractures started at the surface; with glazed rods 80 per cent of fractures were initiated in the interior.

Similar results were obtained by Skarbye (1966). He states that, due to the lower expansion of the glaze, the stress distribution created in the glaze-body system is such that fracture always originates in the interior of the body and not at the surface. The interior of the body withstands a higher stress than the surface layers because the surface is exposed to water vapours which, in normal atmosphere, reduce strength. Therefore, a glazed body is stronger than an unglazed one.

It will be seen from Table 10.1 that hard porcelain is improved by glazing in all its strength aspects, especially in modulus of rupture, which is a measure of surface strength.

Glaze can also influence shape and shrinkage of hard porcelain. Certain glazes low in or free from alkalis cause hard porcelain tiles to warp during firing and prevent the body from fully shrinking although the body is, nevertheless, vitrified; the glaze almost seems able to stretch the body thereby affecting its strength. The difference in shrinkage can be as much as 5 per cent. If such glazes are applied to the outside only of a thin porcelain cup the rim of the cup is drawn in, if applied to the inside only the rim flanges outward.

1.4. *Strength of Various Types of Pottery*

1.4.1. *Feldspathic earthenware*

This is appreciably stronger than the other two, less important, types of earthenware, mainly becuase of its lower porosity. In certain cases it is even stronger than some unglazed hard porcelain, as will be seen from the modulus of rupture figures in Table 10.1.

Endeavours to increase the strength of earthenware have mainly been concerned with the free silica content. Elimination of particles of silica larger than 50 μ resulted in improved strength. Greater, although not spectacular improvements were achieved by eliminating free silica altogether and by replacing it with other fillers (e.g. alumina) of lower thermal expansion and free from inversions.

1.4.2. Vitrified Hotelware

This can be regarded as feldspathic earthenware with little or no apparent porosity. Hence it is much stronger than earthenware. In most vitreous-china bodies there is slight residual apparent porosity and it has been shown that the highest mechanical strength was achieved just before complete vitrification (Harris, 1939, and others).

In the vitreous-china body, contrary to hard porcelain, the free silica crystals originally introduced have not started to go into solution and are preserved in the fired body. Therefore a high crystal/glass ratio is maintained and probably because of this, vitreous china is stronger than hard porcelain. Replacing crystalline silica by alumina and zircon also increases the strength of vitreous china.

1.4.3. Hard Porcelain

Broadly speaking, in both earthenware and vitreous china, the crystals originally introduced by the fillers remain unchanged. In hard porcelain the reactions are carried a stage further to bring some of the free silica into solution to form more glass and thus enhance translucency. The resulting decrease in crystals and increase in glass produces a reduction in strength. Singer (1936) was the first to draw attention to the fact that normal hard porcelain was much more fragile than bone china or vitreous china.

The effect of quartz on the strength of porcelain has been investigated very extensively. Increasing quartz content causes an increase in the strength of hard porcelain, according to some investigators, a decrease according to others. These apparently conflicting results can be explained by the fact that different body compositions were chosen as starting points.

If in a composition high in quartz (e.g. above 30 per cent) the quartz content is raised, the additional quartz crystals are not dissolved in the glass present because this is "saturated" with silica; the resulting increase in the amount of crystals at the expense of glass is responsible for the increased strength.

If, on the other hand, the quartz content is increased in porcelain low in quartz (e.g. 15 per cent) to say 20 or 25 per cent at the expense of clay, the "silica hungry" unsaturated feldspar glass dissolves the added quartz crystals;

the consequent increase in glass and the decrease in mullite crystals, due to less clay, decrease the strength.

On balance, the improvement in strength by increases in free silica in a high quartz porcelain is much less than that achieved by a substantial decrease in free silica in a low-quartz porcelain or by eliminating free silica altogether. Hard porcelain bodies containing little or no free (undissolved) silica are not only much stronger than normal hard porcelain of the classical composition (50 per cent clay substance, 25 per cent feldspar, and 25 per cent quartz) but surpass vitreous china. Scaly crystals of primary mullite, due to the lattice break-up of clay, are thought to be the main source of strength of such porcelains. Secondary mullite (large needle-shaped crystals) formed from the feldspar melt and orientated in random fashion does not contribute to strength.

Improvements in the strength of hard porcelain have also been achieved by crystalline fillers characterized by a high modulus of elasticity, a thermal expansion compatible with those of the remaining constituents of the body, and chemical inertness (insolubility in the glassy matrix). As in the case of earthenware and vitreous china, alumina is best suited for this purpose, but zircon has also been used successfully.

1.4.4. Bone China

Bone china contains approximately 71 per cent crystals (44 per cent beta-tricalcium phosphate and 27 per cent anorthite) and 29 per cent glass. In hard porcelain the ratio is reversed. Moreover, the crystals in bone china are very much smaller than those in porcelain (so small that the resolving power of the optical microscope is insufficient to reveal them). Both these facts are good reasons why bone china should be much stronger than hard porcelain. This is in fact so. The first evidence was provided by Odelberg (1931) when he published the results of tests by the Swedish Government Testing Institute, Stockholm. The summarized values are shown in Table 10.1 (Dinsdale, 1961). The chipping resistance of bone china was over four times greater than that of hard porcelain, modulus of rupture and impact strength twice that of hard porcelain.

Yet bone china is generally regarded as being excessively fragile. This is understandable from the layman's point of view because it looks so delicate; he has at least a psychological reason for expecting bone china to be fragile. However, the notion of fragility of bone china has been perpetuated even by those who are considered authorities in ceramic technology. The well-known ceramist Albers-Schoenberg (1960), 30 years after Odelberg's publication, wrote: "For dinnerware, subjected to daily use, it [bone china] is too delicate. The hard 'Dresden china' is mechanically stronger . . .".

Yet at the factory in Gustavsberg in Sweden, where several types of whiteware are manufactured, bone china was made exclusively for extreme conditions (i.e. use in hotels, hospitals, military and similar institutions) in which highest mechanical strength is required.

1.4.5. *Glass Ceramics*

The high strength of this new type of opaque tableware may be attributed to the high amount of crystals and their small size, to fewer stresses between crystal and glass, to fewer Griffith's cracks and other flaws than in ordinary pottery, and above all to the absence of closed as well as open pores.

2. Thermal Strength

Thermal strength, usually referred to as *"resistance to thermal shock"*, is of prime significance for cooking ware. Man's first pottery vessels were used for cooking.

2.1. *Assessment of Thermal Shock Resistance*

According to the ISO (International Standards Organization) test for thermal shock resistance of *Laboratory Porcelain*, a crucible of the prescribed shape and size is heated at 250°C for 15 minutes and dropped in a water bath of 20°C. If, after this treatment, the crucible is free from cracks or other signs of thermal shock failure, it has passed the test.

In practice, potters devise their own tests, broadly based on the ISO Standard [which, incidentally, is identical with the BSI (British Standards Institution) specification from which it was taken]. Usually more than one test piece (as a rule in the shape of an actual cooking vessel) is used, and each specimen is subjected to, say ten test cycles; sometimes the severity of testing is increased by raising the heating temperature incrementally until failure occurs. The conditions of testing are much fiercer than those encountered in actual use. The temperatures are considerably higher than in the cooker; quenching is occasionally done by placing the vessel on a cold slab or by holding it under a cold water tap.

In line with practical application the vessel often contains water when it is heated. This implies a less severe test than dry heating (which would place vitrified ware under a disadvantage compared with porous one). Water, even when boiling, keeps the vessel cool and thus helps to reduce thermal stresses. In some tests chilling proceeds as soon as the liquid starts to boil, in more severe tests the boiling water is allowed to evaporate completely (thus simulating "dry" heating) before quenching.

The above testing procedures, although eminently practical, are cumbersome and time-consuming. A more elegant method comprises the determination of the degree of glaze compression or tension as revealed under the microscope. Results can be expressed in precise figures which bear a perfect correlation with the lengthy practical tests. Other methods are based on acoustic emission, registering microcracking, and on activation energy which is a measure of thermal sensitivity.

2.2. *Factors Influencing Thermal Shock Resistance*

2.2.1. *Physical Characteristics*

Resistance to thermal shock is largely governed by *thermal expansion*. The reversible thermal expansion is dependent on the bond between the ions or atoms in crystals. Pottery, being made up of crystals consisting entirely of ionic or homopolar bonds, has a low thermal expansion compared with metals. This should be an asset as far as thermal shock resistance is concerned since fewer stresses would be set up. Yet despite its relatively low expansion, most pottery is sensitive to heat shock or sudden cooling and, as with mechanical shock, this is basically due to lack of ductility.

When a vessel is heated rapidly and unevenly the surface expands faster than the interior and is subjected to a compressive stress while the interior is in tension. On cooling the reverse is the case. If at any time the stress set up exceeds the strength (in this case tensile strength) of the material, thermal shock failure occurs. *Mechanical strength* thus determines thermal shock resistance to a certain extent; a low *modulus of elasticity* (*Young's modulus* obtained by measuring *surface strain*) and a high *modulus of rupture* are desirable for good thermal shock resistance. However, mechanically strong bodies do not necessarily show high thermal strength. For instance, bone china, the mechanically strongest type of pottery, cannot be used for cooking because of its high expansion.

High *thermal conductivity* is desirable, but it is a property over which the potter has little control, except with special high-alumina porcelains.

Porosity can be beneficial under certain circumstances. The cooking pots of early man and so-called "primitive" cultures are porous. In the presence of large pores a crack, caused by thermal stress, terminates at a large pore and thus can do no harm. However, the bodies with the highest thermal shock resistance are dense.

With non-porous bodies thermal expansion is much more important. Porous bodies with a relatively high thermal expansion, e.g. earthenware (coefficient $7-8 \times 10^{-6}$) can be used as cooking ware, whereas non-porous bodies with a high expansion, viz. bone china (coefficient $8.5-9 \times 10^{-6}$) cannot.

It should be pointed out, however, that on the rare occasions when bone-china cups or teapots crack on hot or boiling liquids being poured into them this failure is not due to thermal shock. In such cases a minute crack, having been caused by mechanical shock, is present. Such a knock would probably have caused other pottery to crack or even fracture; but with bone china the crack was too small to be seen and was only opened up and made visible by the secondary shock due to sudden heat. Tests at the laboratory of the Worcester Royal Porcelain Co. Ltd. have shown that bone-china cups chilled in a deep-freeze refrigerator, followed by immediately pouring boiling water into them, stood up perfectly well, no matter how often they underwent this drastic thermal shocking. On the other hand, bone-china plates and meat dishes should not be warmed in a cooking oven, especially if flames touch them, because the dry heat in the oven is too fierce, its temperature being far higher than that of boiling water.

2.2.2. Body Microstructure

Dense pottery vessels may show sudden failure after a long time of satisfactory usage. This is due to *"aging"*, similar to *metal fatigue*, i.e. changes in the microstructure, caused by internal stresses which result in permanent and superimposed temporary deformations under repeated heating and cooling (Masson, 1948).

Broadly speaking, the microstructural factors beneficial to mechanical strength of bodies are also desirable for good thermal shock resistance, viz. high crystal/glass ratio, small crystals of low thermal expansion (i.e. mullite) and freedom, as far as possible, from defects.

2.2.3. Design

Design (shape) and thickness of the vessels have an enormous influence on thermal shock resistance. Josiah Wedgwood was aware of this when he stated: "The more uniform the thickness, the thinner and also the more rounded the vessels are, the better they stand up to thermal shocks" (Royston and Barrett, 1958). Sharp edges and corners imply high stress concentration and should, therefore, be avoided. The ideal shape is that of a laboratory porcelain casserole, as Wedgwood has indicated.

2.2.4. Glaze

Good body/glaze match is the most decisive element contributing to satisfactory thermal shock resistance. As pointed out before, the glaze should be under *compression* on cooling. The higher the compression of the glaze, the better the thermal shock resistance. There is, of course, a limit to the optimum degree of glaze compression; if that limit is exceeded, the ensuing excessive tension of the body would cause the article to be disrupted and even shattered (see Chapter 7, section 3.2, p. 126).

A perfect body microstructure, as indicated above, can be made ineffective if the body is covered with a glaze which is under tension or under insufficient compression. On the other hand, a "poor" body, having relatively large—high thermal expansion—quartz crystals, a high glass content and an unsatisfactory texture (i.e. defects) may produce good thermal shock resistance if coated with a suitable, low thermal expansion glaze. In fact, thermal performance may be improved by deliberately increasing the thermal expansion of the body, as in earthenware, through increasing the amount and size of quartz crystals, to widen the thermal expansion gap of body and glaze and thus induce enhanced compression of the glaze.

It is reiterated that a low thermal expansion of the glaze is achieved by a strongly bonded structure, brought about by high silica and alumina contents and a high alkaline earth/alkali ratio.

2.3. Oven-proof and Flame-proof Ware

So-called *"oven-proof"* cooking ware can be made of porous types of pottery within the range, comprising "common" pottery and fine earthenware, as well as of dense bodies like stoneware and porcelain. Replacement of free silica by alumina or bauxite in vitrified earthenware body compositions has resulted in bodies of lower thermal expansions which are suitable for cooking ware, provided they are glazed with a low expansion glaze which, necessarily, requires a higher firing temperature or a firing in which body and glaze are matured simultaneously, as with hard porcelain.

For truly *"flame-proof"* ware, only sophisticated pottery, like cordierite porcelain, lithia-based bodies and glass ceramics of very low thermal expansion, are suitable. The glazing difficulties of such materials are recalled (see Chapter 9, sections 1.4, 2.3, and 2.4). The poor glaze quality is such that foodstuff sticks to the vessel and it is extremely difficult or impossible to remove sticking marks. Such cooking ware is, therefore, not popular. The reader is again reminded that man's first pottery vessels (mainly made by women) were used for cooking; they were placed on a "real" fire and were thus truly "flame-proof".

3. Density

Pottery has a lower density than metals and it was the lightness of porcelain plates that, amongst other things, appealed so much to members of the aristocracy of the seventeenth and eighteenth centuries that they had their plate melted to pay for the new "white gold".

The specific gravity of the great majority of fired bodies, as well as many of the raw materials used, is around 2.5. A similar figure applies to glazes, excepting lead glazes, which, of course, have higher specific gravites. Bone china is a little heavier (specific gravity about 2.8) than other vitrified ware, due to the calcium phosphate having a higher specific gravity than clay, feldspar, quartz, etc. The more porous the body the lower the density and the lighter the article.

The potter distinguishes two kinds of porosity: *Apparent porosity* is the pore volume accessible to liquids under normal conditions (i.e. pressure). *True porosity* is the sum of open and closed pores. At high porosity all pores are accessible, so that there is no difference between true and apparent porosity.

4. Behaviour under Microwaves

Microwaves pass through pottery without affecting it in any way. It is, therefore, a suitable vessel material for cooking in microwave ovens in which it suffers no thermal stress. Bone china, like other dense pottery, can thus be safely placed in microwave ovens. Porous (*unglazed*) pottery is not recommended for microwave cooking. Certain types of glass ceramics have been known to explode in microwave ovens.

It is important to exclude metallic handles and metal decoration, viz. gold or platinum bands, as they are liable to cause arcing under microwaves; this could harm not only the pottery vessel but damage the magnetron of the microwave oven itself.

5. Optical Properties

5.1. *Colour*

The colour of pottery (disregarding decorating) is governed by the impurities present in the raw materials. To reiterate, the most common inpurity is iron oxide; its effect is enhanced by titania, which in itself is white. Iron provides all shades of brown, ranging from yellow to deep red, and sometimes produces greens if fired in a reducing atmosphere, e.g. in the famous *celadon* glazes on Chinese porcelain.

Throughout the ages there has been an endeavour to purify pottery, a striving towards whiteness. In some cases this has been achieved by engobes or white opaque glazes. The main factor which made the production of white bodies possible was the discovery of china clay. The greater whiteness of hard porcelain compared with fine earthenware or vitreous china is due to the iron oxide, invariably contained in the raw materials (even if only in very small amounts) being present in the reduced form (ferrous oxide), which has a much lower staining effect than the ferric form. Even so, hard porcelain looks grey or bluish by the side of bone china, the "whitest" of all pottery. The extra whiteness of bone china used to be ascribed to the bleaching effect of the lime present in it; the phosphate content is now thought to be the chief cause of the exceptionally white colour, the higher oxide of iron forming colourless compounds with phosphoric acid (Weyl, 1941).

5.2. *Characteristics Relating to Translucency*

Translucency, the potter's highest aesthetic achievement, distinguishes the porcelains from all types of pottery.

5.2.1. Factors Influencing Translucency

If crystals are immersed in a liquid of the same refractive index they become invisible. This is due to the fact that the path of light is not broken. Such conditions prevail in clear glass which is, therefore, transparent. So are many crystals and, of course, air. Porcelain is composed of glass, crystals, and voids (air), and is not transparent because glass, crystals, and air have different refractive indices. It is, however, translucent. The smaller the differences in refractive indices of the constituent phases of the porcelain the greater the translucency. The refractive indices of the constituents of fired bone china and hard porcelain are as follows:

	Bone china	Hard porcelain
Glass	1.50	1.48
Quartz	—	1.52
Mullite	—	1.64
Anorthite	1.58	—
Beta-tricalcium phosphate	1.62	—
Range	0.12	0.16

The glassy phase is the main contributor to translucency. High glass content can be brought about by: (a) incorporating a glass itself as one of the raw materials of the body, as in the early frit porcelains and Belleek china, or (b) reaction of the constituent materials, as in bone china and hard porcelain.

A high glass content is achieved by firing the porcelain 150°C higher than its vitrification temperature, and by increasing the amounts of feldspar and quartz. The type of quartz used may materially affect translucency; one which inverts quickly to cristobalite during firing dissolves just as fast in the liquid phase of the porcelain body, thus enhancing translucency (Schüller, 1979) through increased glass formation. High clay content leads to the formation of mullite crystals which reduce translucency. Crystal content is, however, not necessarily detrimental to translucency. (Certain technical ceramics made of pure single oxides or fluorides, consisting entirely of crystals, are as translucent as traditional porcelains). Bone china, which contains more than twice the amount of crystals (and less than half the amount of glass) as normal hard porcelain, is at least equally as translucent. This is due to the smaller difference in refractive indices of crystals and glass present in bone china, and possibly also to the small size of the crystals.

Broadly speaking, complete vitrification is a prerequisite of translucency, although certain pottery bodies of slight apparent porosity are somewhat translucent. Earthenware, in spite of containing possibly more glass than bone china, is not translucent because of its network of interconnected pores. These are filled with air whose refractive index varies greatly from those of glass and crystals.

A few closed pores are always present in conventionally made vitreous pottery. If they could be eliminated translucency would be much improved.

It is possible to remove closed pores in polycrystalline ceramics by special atmospheric control during firing. *Lucalox*, a pure alumina ceramic produced in this way, is almost transparent.

The true porosity of bone china and hard porcelain is increased by over-firing. One would expect that this would reduce translucency, but this is not so. This apparent paradox can be explained by the fact that the larger voids, produced on over-firing, absorb the smaller ones; therefore the number of pores decreases so that the path of light is broken less often—hence greater translucency. In the case of hard porcelain, over-firing causes more glass to be formed, which is a contributory factor to enhanced translucency.

Translucency is, of course, reduced with increasing body thickness. However, the rate of this decrease varies with different types of porcelain. The glaze sometimes increases the transluency despite increasing the thickness of the piece.

5.2.2. A New Assessment of the Aesthetics of Porcelain

Translucency is measured by the amount of incident light being transmitted through the translucent body. However, transmittance is not the only factor involved. Where light is incident upon the surface of a solid, that light may also be absorbed in the surface layer of the solid and, moreover, it may be reflected back from the surface, apart from being transmitted.

Dinsdale (1976) recognized that transmittance was not the only parameter relevant to expressing the aesthetic quality of translucent ceramics; reflectance was of equal, indeed greater, importance. Transmittance and reflectance of different types of porcelains are plotted in Fig. 10.3. It will be seen that American porcelains ("household chinas") have high transmittance but low reflectance, that Continental European hard porcelains cover a wide range (high reflectance not coinciding with high transmittance), but that all bone chinas, wherever made, fall within a small, closely limited, area of the chart, denoting maximum reflectance and high transmittance; this explains the exceptional beauty of bone china.

6. Mechanical Wear Resistance

Resistance to mechanical wear concerns the surface of articles and, therefore, only affects the glaze.

6.1. Hardness

6.1.1. Assessment of Hardness

Surface properties, such as resistance to abrasion, often referred to as "hardness" or more specifically *"scratch hardness"*, are notoriously difficult to measure with any degree of accuracy.

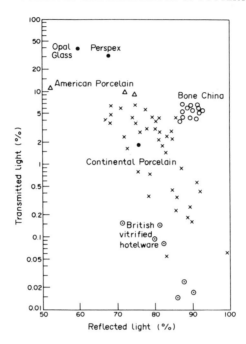

FIG. 10.3. Transmittance-reflectance plot of various translucent bodies. (After Dinsdale (1976) p. 976.) Reproduced by permission of The American Ceramic Society.

The well-known "Mohs" scale is not sufficiently sensitive to differentiate between pottery glazes; these fall between 5 and 7 of that scale, the majority between 6 and 7.

Diamond pyramid hardness test is more sensitive than Mohs hardness but not always easy to interpret. A four-sided diamond pyramid indenter is brought in contact with the test piece under controlled load. The size of the indentations of a diamond pyramid, applied to the test piece, is a measure of hardness. The diamond pyramid hardness can be applied to minute volumes of glaze (or glass) the spot being chosen by the microscope, which forms part of the equipment. Any flaw can, therefore, be avoided, and a true assessment of the glaze surface obtained.

6.1.2. *Aspects of Mechanical Wear*

Cutlery is softer than glazes and it is often difficult or even impossible to produce a scratch mark with a sharp steel knife, even on the comparatively soft vitreous-china or bone-china glaze. If scratches are difficult or impossible to produce at will they nevertheless do occur inadvertently. The scratching action is enhanced on wet surfaces.

This also applies to the so-called "scuffing"—the most frequent cause of wear—which happens during washing up, when one plate is allowed to slide on to another, or when plates are just stacked. Glazes of identical hardness, especially those for low temperatures, scratch each other. However, hard glazes, like those of hard porcelain, harm each other less than do soft glazes. Attempts have been made to coat the backs of plates with a softer glaze than the front so that the back of one plate does not mark the front of another, particularly during washing up. Efforts to increase the wear resistance of pottery glazes by organic coatings have not generally met with success.

6.1.3. Factors Influencing the Hardness of Glazes

The more strongly bonded the glaze, the better the scratch hardness and abrasion resistance. High bond strength of glazes is achieved by high contents of the glass network-former silica and he glass co-former alumina. Such glazes, incidentally, have low thermal expansions and high chemical resistance (important for laboratory porcelain) and require high firing temperatures. All these aspects are incorporated in hard porcelain glazes which are technically superior to all other glazes.

6.2. Metal Marking

Metal marking, more often described as "*silver marking*", although not confined to silver, is caused by certain cutlery producing grey streaks or lines on the glaze which cannot be removed by ordinary rubbing, yielding only to mild abrasive powders. High magnification of the glaze surface reveals that even the best glaze is not smooth but somewhat serrated and undulated. The soft metal (silver) worn away by the high spots of the serrations is deposited in the microscopic hollows of the glaze surface.

7. Chemical Wear Resistance

7.1. The Chemical Inertness of Pottery

The chemical resistance of pottery is extraordinarily high. It is the property responsible for its indestructibility. Pottery, by the very fact that it consists of oxides, is not liable to oxidation; it is resistant to humidity, rain water and to chemicals in the soil; it is not attacked by acids, not even by the strong mineral acids, such as hydrochloric, sulphuric and nitric acids. (An exception is the bone china body, the phosphate constituents of which are soluble in hydrochloric acid; this is of no practical significance as the body is covered by the glaze which withstands the attack.) Only hydrofluoric and phosphoric acids are harmful to pottery, as are corrosive alkalis.

7.2. *The Effects of Detergents*

Most pottery is glazed. Therefore, chemical (like mechanical) wear resistance does not affect the body. Like bodies, glazes on all types of pottery, with very few exceptions, are chemically inert, as far as normal household chemicals, including detergents, are concerned. However, detergents can be harmful to on-glaze colours (some of which, like blues based on cobalt, may also be affected by vinegar and fruit juices). On-glaze low-temperature stains, consisting of pigment (metal oxide) and flux, have weakly bonded structures and are, therefore, liable to be attacked. Resistance to detergents is not only influenced by the flux but also by the pigment. Moreover, the composition of the underlying glaze is important; the more strongly bonded the glaze, the greater the resistance of the stain. Even the body is capable of influencing the resistance to detergents of on-glaze colours through ion migration from the body to the surface of the glaze.

7.2.1. *The Influence of Detergent Application*

The *temperature* of the washing-up liquid overrides all other factors; the higher the temperature, the worse the attack, severe damage to on-glaze decoration occurring at 70°C.

Washing in *automatic dishwashing machines* is far more harmful than *manual washing*, because the temperatures in the dishwashers are far higher than would be tolerated by the bare skin and because the ware is exposed to the washing up liquid far longer. Many dishwashers incorporate a programme for "delicate" pieces, implying lower temperature and shorter duration of the washing cycle. Dishwashers incorporating two washes separated by a rinse are preferable for on-glaze decoration. Even with these precautions no crockery with on-glaze decoration can be described as "dishwasher-proof".

The *type of detergent* is of considerable significance, although it is not nearly so decisive as temperature. Most modern detergents are complex phosphates which are less fierce than the formerly used partly corrosive alkalis. Manufacturers of detergents sometimes change the composition of one of their products without changing its name. Thus, a harmless washing agent may suddenly become harmful.

The *concentration* of the detergent solution has only a marginal effect on on-glaze decoration in respect of manual washing. As far as dishwashers are concerned, the amount of detergent is prescribed; an insufficient amount may cause the formation of white films on the ware; these give the impression of bad fading, but they can be removed by manual washing. The correct amount of detergent depends on the hardness of the water; it should be reduced in the case of soft or softened water.

7.2.2. *Methods of Testing*

With the customary laboratory tests, the specimens, usually in the form of actual ware, are immersed in the standard alkali solution or detergent solution under standard conditions of concentration, temperature and duration of immersion. The first indication of wear is loss of gloss; later manifestations are fading of colour.

These laboratory methods of testing bear little relationship to the actual conditions in a dishwasher. The water vapour generated in a dishwasher is far more corrosive than the hot immersion solution used in the test. For these reasons straight cycling tests in the automatic dishwasher are carried out in conjunction with laboratory tests. This means that the dishwashing machines have to run continuously until a reasonable number of washes are simulated. Nevertheless, such tests take too long if quick results are required.

A system, using a spectral photometer which measures complementary wavelengths, gives results in precise figures (Lohmeyer, 1968, 1969). The measurement of the loss of gloss of on-glaze colours, quantifying dishwasher detergent attack on the colours (Cubbon and others, 1984), can be regarded as the most expedient and meaningful type of test.

7.3. *Release of Lead and Other Toxic Metals*

7.3.1. *The Problem*

Lead, in one form or another, has been used for a very long time as an exceedingly useful flux in glazes and stains. *Red lead* (lead oxide PbO) was commonly used but, being toxic, it caused lead poisoning to the glaze and colour worker. Fritting the toxic oxide with silica produced a non-toxic glass which still forms the main constituent of most glazes and stains (see Chapter 7, section 7.1., p. 137). The health of the glaze and colour workers has thus been protected. Lead poisoning in the pottery factory has become a thing of the past.

However, in the nineteen-sixties one or two cases of severe lead poisoning, even death, came to light (this time involving the user of the finished pottery product). It was traced to drinking fruit juice which, for a long time, had been stored in vessels made by peasant potters, and to imbibing wine which had been boiled in such pots.

7.3.2. *Legislation to Make Pottery Chemically Safe for the User*

In view of the general pollution of our environment, particularly by lead fumes from car exhausts, health authorities throughout the world have become more vigilant. Bearing in mind the unfortunate occurrences of lead poisoning mentioned above, legislation was established, first in the United States, later by many other countries, following the American lead in the hope of preventing such accidents and making pottery safe for the user. All pottery had to pass certain tests before it could be placed on the market. Legislators were mainly concerned with imported articles. There have, indeed, been cases where foreign consignments were refused importation by several countries.

It was considered that immersion of the ware in 4 per cent acetic acid at 20–25°C for 24 hours and determination of the amount of lead released by atomic absorption spectrophotometer, represented the most expedient and foolproof test; it has been adopted almost universally. The limit of permissible release varies from country to country. It is expressed in parts per million (the permissible limit being generally less than 10 ppm) or in mg/dm^2.

In addition to lead, nearly all specifications stipulate *cadmium* as well, an element specifically used in the beautiful selenium reds and much more toxic than lead; its limit, as a rule, is 10 per cent of that of lead.

7.3.3. *Factors Influencing Lead Release*

The main criterion in lead release is again the degree of *bond strength*, not only of the stain but also that of the underlying glaze. Loosely bonded structures show high lead release.

Lead release is not necessarily the greater the higher the lead content of the stain. For instance, in a lead–boron–silica system increases in lead with corresponding decreases in boron result in *less* lead release because of the stronger bond of high lead/low boron compounds. Within certain compositional ranges phase separation occurs, a low durability lead boron phase and a silica-rich phase being formed. The addition of alumina represses phase separation and hence reduces lead release (Takashima, 1985).

Under certain circumstances the *pigment* can have a vital influence on the bond structure and hence lead release. Perfectly safe, transparent lead glazes, releasing less than 1 ppm lead, are transformed into lethally weak structures by the addition of copper, resulting in a lead release of hundred times the permissible level. Such beautiful, but deadly, copper glazes were responsible for the cases of lead poisoning mentioned above. Since then it was found possible to suppress lead release in copper-bearing glazes by $BaCrO_4$ and more effectively by chromium sequioxide Cr_2O_3 (Quon and Bell, 1980).

Higher *firing temperatures* of the stain and also of the ware itself bring about a stronger bond and hence reduced lead release.

The *atmosphere* in the decorating firing has an enormous influence. A static atmosphere produces worse lead release (Eppler, 1977). Humidity greatly reduces it. Gas-fired kilns in which steam is generated naturally are, therefore, preferable. In electric kilns the installation of an auxiliary gas burner will achieve the desired steam effect (Roberts and Holmes, 1982).

Cadmium release from selenium reds can be prevented by *encapsulating* cadmium sulphoselenide pigments in zircon (Roberts and Holmes, 1982).

7.3.4. *Extraction of Lead from Ceramic Tableware by Foodstuffs*

A number of research workers have studied the correlation between the statutory tests for lead release from pottery and the actual amounts of lead passed from pottery vessels into food for human consumption. Cubbon and others (1981) obtained the following main results:

1. The amount of lead extracted was dependent on the pH and the temperature of the foodstuffs, confirming previous findings by Frey and Scholze (1979).
2. The amount of lead extracted fell very rapidly on successive treatments to a virtually constant level.
3. Excepting most acidic foodstuffs, the amount of lead extracted was practically independent of the acetic acid release test value in the range of 1 mg Pb/l to 14 mg Pb/l.
4. The removal of lead from the glaze and decoration surface in the first few treatments appeared to increase the percentage of silica content of the surface of the test plates and to reduce the susceptibility of the surface to acid attack.
5. Measurements allowed the conclusion that the extraction level of acid foods of 2–4 mg lead per meal persisted for virtually all the lifetime of the ware. The lead intake from pottery per person per day was thus unlikely to exceed 10 μg lead, compared with 100–300 μg lead which, according to published literature, was the dietary intake per person per day.

A very striking example of the minimal intake of lead was provided by Bernhard (1974). He examined a salad bowl which showed a lead release as high as 1000 ppm; the limit prescribed by DIN (Federal Republic of Germany) legislation was 6 ppm; yet 1 litre of potato salad, after having been kept in that bowl for 1 hour, contained only 0.24 ppm of lead.

Conditions are more severe with *cooking ware* than with normal tableware, as foods (some of acidic nature) are held at higher tmeperatures, and for longer periods. For this reason, regulations concerning release from pottery cooking vessels are more stringent. Cooking ware should not have on-glaze decoration on the inside and should have a strongly bonded glaze without lead or boron which implies that it should be made of stoneware or porcelain (for the sake of low lead release, as well as good thermal shock resistance).

A study concerned with the effects of foodstuffs on the durability of glazes (with special potential colour effects) containing natural *thorium* and natural or depleted *uranium* revealed that the potential hazards to users would be negligible, even under the most stringent requirements (Stradling and others, 1972).

7.3.5. *Conclusions*

Two questions arise:

1. Are the tests of lead release from pottery sufficiently stringent and are the limits of permissible lead release adequate to ensure 100 per cent protection to the health of the user?
2. Will the established legislation thus prevent any recurrence of lead poisoning to the user of pottery vessels, worldwide?

The answer to the first question is a resounding "Yes", that to the second one an equally resounding "No".

When legislation was first introduced a large range of articles, made by both industrial and studio potters, did not conform to the limits laid down. Radical measures were required. The suppliers of colours and glazes have ensured that—by the means indicated in section 7.3.3., viz. improving the bond strength—their products containing lead were safe. As an extra safeguard many pottery manufacturers introduced steam into the atmosphere of the decorating firings.

Release figures, well below those permissible, have been achieved as a result of the steps taken. The correlation studies mentioned in section 7.3.4 have shown that only a small fraction of the lead released is transferred from pottery articles into the actual food, the amount varying with different types of foodstuffs. One particular example, cited in 1974, thus before the measures of preventing lead release had taken full effect, highlighted this point. A vessel which released more than 160 times the amount of lead permissible, transferred only one-twenty-fifth of the allowable lead to the foodstuff. Moreover, the daily intake of lead from foostuff itself (not in contact with pottery) is 10–30 times greater than the daily amount of lead passed from pottery into the food.

Yet, despite these reassuring facts, somewhere on this earth peasant potters, perfectly honest and honourable people who, however, may derive their materials from doubtful sources, produce beautiful but—unknown to them—lethal wares with an exquisite low-temperature copper glaze. How are unsuspecting customers to know that they must not store fruit juices in these lovely vessels or use them for boiling wine?

The powers that be have gone to almost absurd lengths to make pottery safe, but so far they have seemed to be powerless in preventing fatalities outside the net of legislation. The only consolation lies in their infinitesimally small number.

Further Reading

Mechanical strength:
    General: Astbury (1966); Binns (1962); Budworth (1970); **Dinsdale and Wilkinson (1966)**; **Dinsdale and others (1967)**; Dinsdale (1986); Kerkhof (1983).
    Work of fracture: Clarke and others (1966).
    Effect of geometry: Wade and Borty (1966).
    Strength of porcelain: Dietzel (1953); **Masson (1956, 1962)**; **Schüller (1962, 1967)**; **Skarbye (1964, 1966)**; Wiedemann (1966).
Thermal properties:
    Specific heat capacity: Dinsdale (1986, p. 222).
    Thermal shock theory: **Dinsdale (1986, p. 221)**; Royston and Barrett (1956).
    Thermal expansion: Schüller and Lindl (1964).
    Moisture expansion: Vaughan and Dinsdale (1962).
Translucency: **Dinsdale (1976)**; Dinsdale (1986, p. 236).
Glaze hardness test: Ainsworth (1956); Roberts (1965).
Abradability of glazes: Molnar and Wagner (1974).
Chemical resistance:
    Effect of detergents: Bernhard (1976); Lohmeyer (1968, 1969); **Cubbon and others (1984)**; Rado (1987).
    Lead release: Bernhard (1974); **Cubbon and others (1981)**; Frey and Scholze (1979); Krajewski and Ravaglioli (1980); **Roberts and Holmes (1982)**; Takashima (1985).
General durability: Rado (1986).

11

The Pottery Industry

The pottery craft marked the beginning of civilization and is therefore as old as civilized man himself. The pottery industry, rooted in the craft, is traditional, but this does not necessarily mean that it is backward. Pottery today, despite competition from newer materials, is holding its own and is becoming a highly efficient industry.

1. The Past

Pottery was perhaps the first craft to be transformed to an industry. This happened in the eighteenth century, the great age of experimentation, particularly in Britain. Pottery then stood at the forefront of technological development and as the most progressive industry it occupied a position comparable to the electronics or aircraft industries of today. (Potters themselves seem to have forgotten this and it was left to an eminent lord of the realm to remind them that, in England, the Industrial Revolution began in the potteries.)

Josiah Wedgwood and Josiah Spode, the great potters of the golden age of pottery, not only excelled in their own chosen industry, they left their mark on the technological and economic growth of the country as a whole. Wedgwood not only rationalized pottery and raised it from a craft to an industry, he also initiated division of labour, improved the living standard of his workmen, and applied scientific methods in his experimentation. Furthermore, he introduced modern business methods and corresponded with his foreign customers in their own language, he sought the help of attachés at British embassies in foreign lands to promote trade; in his efforts to further industrial progress and commerce generally he built canals. Spode, the originator of bone china, improved roads and introduced a "fire-engine" long before Watt's steam engine came on to the scene.

The nineteenth century was a period of consolidation, it was also a time when scientists like Seger (1902) first studied pottery. But the pottery industry itself did not experience the same uplift as in the eighteenth century and it settled down to a self-contained, somewhat snug life. The craftsman dominated the industry, sometimes tending to distrust the scientist, an attitude which was taken over into the early twentieth century.

Like other traditional industries, the pottery industries congregated in certan centres: East Liverpool in Ohio, in the United States, Selb in Bavaria, Germany, Limoges in France, and Stoke-on-Trent in England. The two latter centres had had a tradition of pottery making long before it developed into an industry. It is significant that these centres, and indeed the countries in which they are situated, have remained loyal to their original products: the United States to vitreous china, Britain to earthenware and bone china, France and Germany to hard porcelain. This shows how strong tradition is in these countries. Where the traditional element is apparently less developed, as in Italy and Scandinavia, feldspathic earthenware, vitreous china, and hard porcelain have been made side by side for some time.

12. The Present

2.1. *Health*

Tremendous strides have been made in eradicating health risks. Before the substitution, in the nineteen-thirties, of flint by alumina as a placing material, the average mortality age of a bone china kiln placer was 30 (Bailey, 1986). Pneumoconiosis, the cause of death of pottery workers (handling pulverized silica materials) at such a young age, has been wiped out by appropriate legislation, as has lead poisoning of operatives engaged in glaze and ceramic colour work. Likewise, asbestos is no longer a health hazard as far as kiln workers are concerned.

Machines incorporate safety devices so that accidents are virtually impossible. Special guidelines on health and accident prevention have been published by The Institute of Ceramics, aimed not only at the industrial potter but also the studio potter (*Health and Safety in Ceramics—A Guide for Educational Workshops and Studios*, Anon., 1986). Reference should be made to the pioneering work by British Ceramic Research Limited (formerly British Ceramic Research Association) on protective overalls and fettling hoods, to reduce the incidence of pneumoconiosis in the pottery industry. This work (Bloor and Dinsdale, 1962) is embodied in the U.K. Pottery Health and Welfare Special Regulations 1950.

In the planning of the layout of new pottery factories special emphasis is placed on the improvement of the environment, viz. a substantial decrease in the emission of harmful gases (partly as a direct result of reduced fuel requirements in the firing) and the elimination of effluent from filter presses (due to filter presses becoming redundant as a consequence of changing over to dry isostatic pressing). Moreover, kilns are designed in such a way that kiln workers are not exposed to abnormal heat, all heavy work having been eliminated (Anon., 1984).

Stringent regulations have ensured that the user of domestic ware is completely protected from any health hazards arising from the release of lead or other toxic metals from pottery glazes and colours. The amounts of these substances, given off pottery and passed into food, are far less than those contained in the food itself. The number of cases of lead poisoning caused by the loophole that legislation cannot always reach the peasant potter, wherever she or he may be, is infinitesimal.

2.2. *Processing*

Up to about the middle of this century very little had changed basically since Wedgwood first established his then revolutionary production units 200 years ago. Within the last three decades the pottery industry has undergone another revolution during which there have been separate stages of development.

Manual processing has given way to machines. This vital step necessitated better controlled and more uniform raw materials. If, in the past, there were fluctuations in the plasticity of the clay, the potter instinctively adjusted his jiggering or jollying operation by varying the pressure he applied by hand or by altering the amount of water he used. The machine is incapable of making the necessary adjustments on its own accord.

Likewise, if there were variations in the alkali content of the fluxing materials, the fireman of the old bottle ovens was able to judge from his "proofs" how he had to adjust the duration of the firing to achieve the desired quality of the fired product. Under present conditions of firing in tunnel kilns with their set programmes the effect of the fluctuations in the flux content of, say, a bone china body would only be revealed after the ware has left the tunnel—when it would be too late; nothing could be done bar a postmortem examination.

It is, therefore, vital that raw materials are consistent. The earth, however, does not always yield non-varying materials and special refining methods, like froth flotation and blending, had to be introduced to ensure consistency.

The rate of development has not been the same throughout the pottery industry. It has been fastest in the cheaper end, viz. earthenware. Changes have been slower in the bone china sector. In the nineteen-sixties there were still ninety different skills in the bone china industry. This diversity of skills was highlighted by the history of a cup, as shown in the flowsheet in Fig. 11.1 (starting where the flowsheets illustrated in Figs. 3.2 and 3.3 leave off (see Chapter 3, pp. 44 and 45). The list of processes is impressive enough without further emphasis. For the sake of clarity, Fig. 11.1 does not incorporate all steps necessary. For instance, it does not include:

(a) The placing for the glost and decorating firings and the emptying.
(b) The inspection after the various operations.
(c) Any repair jobs, e.g. the grinding and polishing after the firing to remove kiln dirt, etc., followed by refiring.

The processing of figures of rich ornamental ware is even more complicated, sometimes involving up to ten decorating firings.

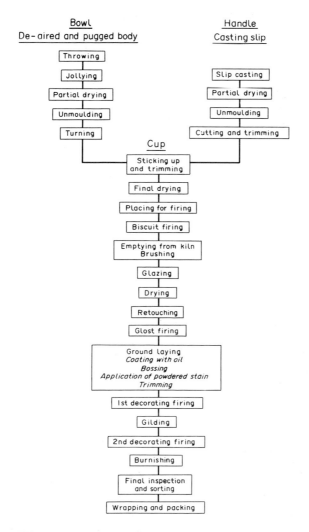

FIG. 11.1.   The history of a bone china cup, flow diagram.

Processing used to be based on a strict division of the various production steps. Between these procedures the ware used to be stored. It has been realized that storage is undesirable because:

(a) it involves extra handling time,
(b) it takes up valuable space,
(c) it causes ware to collect dust, and
(d) it tends to lead to excessive breakage.

By eliminating or at least minimizing storage, the first step to in-line production had been created. However, there were apparently unsurmountable obstacles to manufacturing pottery in a fully conveyorized system, the most important arising from the "memory" of the clay.

The present author remembers having been shocked when, back in the early nineteen-sixties, his boss, Joseph Gimson, a ceramic industrialist of vision, postulated that "clay was the greatest enemy of the potter, all the difficulties and failures of making pottery having their roots in the clay". Clay, the very basis of pottery, being an enemy of the potter? Never! Yet, Gimson's seemingly outrageous statement proved prophetic. Following his (and one or two others) persistent recommendations, dry pressing of flatware became a reality. The absence of water meant that (although clay as such was still contained in the body) plasticity, the very property which, hitherto, had made pottery manufacture possible, was by-passed and removed from the shaping process, water having been the prerequisite of utilizing this property. Thus the memory of the clay could no longer make itself felt and consequently its accompanying manifestations, viz. crookedness and cracking, were expected to disappear.

As a spin-off, dry pressing was instrumental in eliminating the intricate drying process, thus paving the way to fully automatic conveyorized production.

One of the great assets of pottery is its diversity. Dry pressing, or more correctly isostatic dry pressing, has so far been confined to simple shapes like plates and bowls. More complicated shapes are still made by methods relying on the special characteristics of clay: plasticity and castability. These processes are designed with a view to easy integration into automated in-line systems.

Besides the actual shaping, other manufacturing operations like glazing, decorating and the transitional steps of placing ware in kilns, stacking saggars and setters, their emptying, etc., have been mechanized and automated.

Remarkable advances have been made in the development of kilns. The second half of this century has seen the demise of bottle ovens. The inflexible tunnel kilns which replaced the bottle ovens have themselves been superseded by low thermal mass fibre kilns with their recuperative, high-velocity burners, guaranteeing uniformity of temperature throughout the setting and allowing firing times to be reduced from more than 100 hours to less than 100 minutes. Fast firing has saved fuel, facilitated conveyor production, avoided unsocial working hours and been indirectly responsible for making decoration dishwasher-proof.

A number of improved fabrication methods have entailed the use of computers and microprocessors: i.e. in the blending of china clays, in the body preparation and in various process controls, especially in conveyorized systems; even a computer controlled decorating machine has been developed.

3. The Future

3.1. *Availability of Raw Materials*

As pottery is made from materials present in abundance on the crust of the earth, its future seems assured. More intensive refining methods may have to be introduced if and when high-purity deposits become depleted.

3.2. *Energy Savings Through Low-Temperature Bodies*

The more widespread use of fast firing will reduce the amount of fuel required. A more drastic saving in energy could be achieved by adopting bodies which require much lower firing temperatures than do existing types of pottery.

Vitrified bodies have been developed for 1000°C and less, by incorporating low melting fluxes like boron phosphate, or by mixtures of perlite (a siliceous volcanic ash), clay and talc, or by adding a soda/magnesia mixture to china clay and quartz, by melting a mixture of lead bisilicate, china clay and alumina around 1350°C. The frit thus formed produces a dense crystalline body at 870°C. A porcelaneous body of high translucency can be made by impregnating highly porous earthenware with a fluid glass at a low temperature.

These are just a few examples of bodies which, because of their low firing temperatures, will more than halve energy requirements. The fact that such bodies may not contain enough clay for the traditional methods of shaping in the plastic state does no longer matter as dry isostatic pressing, the shaping method of the future (not only for flatware as at present, but eventually for more complex articles), does not require plasticity.

The only problem of clay-free or low clay bodies may be the lack of mechanical strength in the early stages of firing. In traditional bodies strength is maintained through the bonding structure of kaolinite until about 500°C, when its water of constitution starts to be eliminated. At that temperature the forces of sintering take over to provide the necessary degree of strength. With non-clay bodies there are no bonding forces up to 500°C, even if organic binders have been added, because organics are likely to have been eliminated well below 300°C.

Finally, bearing in mind that all domestic pottery is glazed, it would be futile to regard the saving of energy, by lowering the firing temperature, merely as a matter concerning the body. The thermal expansion of the glaze must always be lower than that of the body; the lower the expansion of the glaze, the higher the firing temperature required. The highest temperature possible for a glaze is that required for maturing the body. This implies once firing—a natural technolgoical spin-off of low-temperature bodies.

The sequence of developing low-temperature body/glaze systems would be:

1. Establish the firing temperature of the body.
2. Obtain a glaze of the lowest thermal expansion for that temperature.
3. Adjust the body composition so that the thermal expansion of the body is appreciably higher than that of the glaze.

3.3. *Consolidation*

The industrial potter's ideal is a single machine into which are fed the powdered raw materials at one end and which turns out the fully finished pieces of ware, ready for despatch, at the other end. We are now, in 1987, very much nearer to such a continuous production line than we were 20 years ago.

More pottery factories will go over to dry isostatic pressing. The process of combined shaping and colour decorating (under-glaze) in one single operation (Anon., 1984) will be more widely adopted. There will be an increase in "once-fired" ware. More fast fire kilns will be established, possibly batteries of several intermittently fired tunnels to compensate for the shorter time involved in the preceding processes of shaping and glazing, compared with firing. The various production steps will be amalgamated and replaced by individual production units, possibly one for each type of article. Pottery is likely to change from a labour-intensive to a capital-intensive industry.

The use of computers and microprocessors will be intensified at all production stages. Managers will be able to observe what is happening, at any time anywhere in the factory on a VDU screen (Norris, 1981). The studio potter may use equipment considered to be the microprocessor's equivalent to the "potter's thumb".

As with other industries, microprocessor technology will eliminate more jobs than it will create. This raises ethical and moral questions. The choices are:

(a) Pursuing the ecological approach, implying a reduction in the standard of living.
(b) Applying microprocessor technology 100 per cent, disregarding unemployment.
(c) Doing nothing—the worst choice.

Notwithstanding the desirability of the first choice, it is almost inevitable that the second choice will be the one to be adopted. Nevertheless, the pottery technologist will have some reason to be gratified, as there will always remain the *objets d'art*. Their manufacture cannot be mechanized and will continue to combine meaningful activity and pleasure in work. An arrangement popular in Scandinavia is likely to be applied more universally in which artist potters have their own studios in large factories, enabling them to devote their whole time to their own creations, unfettered by interference from the factory management who, in their turn, relieve the artist from all unproductive chores of administration, marketing and selling, the latter being done in the artist's name.

3.4. *Competitors of Pottery*

Will pottery survive as the mainstay of tableware? Has industrially made pottery any future? There are those who predicted that all ceramic tableware would be ousted by *plastics*. They have been proved wrong. It is true, plastic tableware does not break as easily as pottery, not because it is stronger (which it is not) but because it is more elastic. However, it has a number of disadvantages, as ascertained in exhaustive tests on the melamine-formalde-hyde type, filled with a cellulose base. It is softer than metal and, unlike pottery, can therefore easily be spoiled by deep scratches with cutlery, the crevices produced harbouring dirt and bacteria; it is porous, absorbing liquids, and cannot be cleaned as efficiently as glazed pottery; it does not resist even moderate heat, being affected at temperatures below 100°C at which pottery plates are heated for self-drying purposes; it cannot, therefore, be made sterile; unlike pottery, it lacks chemical durability and deteriorates with time and when exposed to certain liquids (Koenig, 1952).

For these reasons tableware in plastics would appear to be unhygienic and thus quite unsuitable for mass feeding purposes, the main use to which it is put. Improvements have been made in plastic tableware (and will continue), but in spite of this, plastics have not so far seriously undermined pottery tableware and are not likely to do so in future.

Tableware made of *glass* is perhaps a more serious competitor, since it shares some of the good properties of pottery. However, it can become positively dangerous. Toughened glass which is used for tableware is made mechanically stronger by special heat treatment, ensuring that the surface layer is under strong compression. If this surface layer is damaged, catastrophic failure occurs, the whole piece disintegrating into dust. Tumblers made of toughened glass have been known to explode spontaneously into thousands of small fragments. Such failure never occurs with pottery.

For a time *glass ceramics* were considered as the most serious competitor of pottery, especially in the field of cooking ware and hotelware. However, the threat has not materialized and is unlikely to do so in future. It would appear that glass ceramics are more costly to produce; they lack the diversity and the decorative possibilities of pottery which is sold on eye appeal. Translucency can be produced in glass ceramics as well as in glass. However, although both have very high transmittencies, their reflectances are very low. This is probably the reason why they lack the aesthetic appeal of porcelains, especially of bone china.

While there is a craving for beauty left in man there will always be a demand for pottery, the first synthetic and also the first of the most durable manmade materials, known as ceramics.

Further Reading

Historical:
    Wedgwood and the pottery industry: **Green (1930)**.
    Wedgwood and his times: Wedgwood (1964).
    Spode and his times: Thomas (1933).
    Evolution of porcelain: Rado (1964).
    The importance of pottery: Rado (1970).
    The Potteries: Beaver (1964).
Quick freeze method of preparing raw materials; Schnettler (1967).
Low-temperature bodies: Arlett and others (1963); Norton (1958).
Silicate polymers: German (1965).
Reinforcing ceramics: Fishlock (1966).
Use of computers and microprocessors: Frahme (1982); Schacknies (1983); Bartram and Briggs (1982).
Mathematical programming: Gay (1978).
Planning layout of factories: Thormann (1982).
The pottery industry: **Bailey (1986)**; **Dinsdale (1981)**; Johnson (1985); Matthews (1977); **Norris (1981)**.
Health: Bloor and Dinsdale (1962).

APPENDIX
Calculations in Pottery

1. Body Calculations (See Chapter 3, p. 38)

1.1. *Conversion of Chemical to "Rational" Analysis*

In order to calculate the batch composition to correspond to a set rational body composition it is necessary to know the rational analyses of the various raw materials available.

Feldspathic materials and especially clays invariably contain impurities. Conveniently the "impurities" in clay are feldspar (or mica) and quartz, whereas in feldspathic materials they are quartz and sometimes "clay substance" (as kaolinite is usually referred to in the rational analysis).

Let us consider a china clay (Cl) of the following percentage chemical analysis:

SiO_2	48.60
TiO_2	0.05
Al_2O_3	36.80
Fe_2O_3	0.50
CaO	0.16
MgO	0.25
K_2O	1.18
Na_2O	0.26
Loss on ignition	12.20
	100.00

We assume that the alkalis are present as feldspar.
Feldspar contains the following oxides:

		Molecular weight
Potash feldspar:	K_2O	94.2
	Al_2O_3	102.2
	$6SiO_2$: 6×60.08	360.5
		556.9
Soda feldspar:	Na_2O	62.0
	Al_2O_3	102.2
	$6SiO_2$	360.5
		524.7
Lime feldspar:	CaO	56.1
	Al_2O_3	102.2
	$2SiO_2$	120.2
		278.5
		229

The above clay Cl has a potash content of 1.18. The alumina content in the potash feldspar contained in this clay is therefore $\frac{1.18}{94.2} \times 102.2 = 1.28$ per cent; its silica content is $\frac{1.18}{94.2} \times 360.5 = 4.52$ per cent. The percentage of potash feldspar in clay Cl is thus:

K_2O	1.18
Al_2O_3	1.28
SiO_2	4.52
	6.98

The figure of 6.98 per cent can be checked using the molecular weight of potash feldspar, viz. $\frac{1.18}{94.2} \times 556.9 = 6.98$ per cent.

By similar calculations we obtain the compositions by percentage of the soda feldspar, viz.

Na_2O	0.26
Al_2O_3	0.43
SiO_2	1.51
	2.20

and of the lime feldspar, viz.

CaO	0.16
Al_2O_3	0.29
SiO_2	0.35
	0.80

If we add up the alumina contents of the three feldspars, viz. $1.28+0.43+0.29=2.00$ per cent and deduct this sum from the total alumina content of clay Cl of 36.80 per cent we arrive at the alumina content in the clay substance of clay Cl, viz. $36.8-2.0=34.8$ per cent. Clay substance consists by percentage of:

Al_2O_3	102.2
$2SiO_2$	120.2
$2H_2O$	36.0
	258.4

Therefore the silica content in the clay substance of clay Cl is $\frac{34.8}{102.2} \times 120.6 = 41.00$, and the content of combined water is $\frac{34.8}{102.2} \times 36 = 12.20$, thus:

Al_2O_3	34.8
SiO_2	40.93
H_2O	12.2
	87.93

The figure of 87.93 per cent can be checked by the molecular weight of clay substance, viz. $\frac{34.8}{102.2} \times 258.4 \sim 88.0$ per cent.

If we add up the silica contents of the three feldspars and of the clay substance, viz. $4.52+1.51+0.35+40.93=47.31$ per cent, and deduct this sum from the total silica content of the clay of 48.60 per cent, we obtain the content of free silica (quartz) of the clay, viz. 48.60 per cent-47.31 per cent$=$ 1.29 per cent. We thus have by percentage:

Potash feldspar	6.98
Soda feldspar	2.20
Lime feldspar	0.80
Total feldspar	9.98
Clay substance	87.93
Quartz	1.29
Impurities: TiO_2	0.05
Fe_2O_3	0.50
MgO	0.25
	100.00

In practice the sum of the constituents is not always 100 per cent, e.g. if the clay contains disordered kaolinite. The alkalis often form part of mica (instead of feldspar), especially in British china clays and ball clays. The mica formula (i.e. $K_2O.3Al_2O_3.6SiO_2.2H_2O$) should then be applied instead of the feldspar formula.

1.2. Calculation of Batch Composition Based on a Given Rational Analysis

We want to compound a body of the classical rational composition of hard porcelain: 50 per cent clay substance, 25 per cent feldspar, 25 per cent quartz.

Let us assume raw materials of the following rational analyses are available for making up this composition:

	China clay Cl	Feldspar Fl	Quartz sand Ql
	(%)	(%)	(%)
Clay substance	87	0	0
"Ideal" feldspar	11	95	0
Quartz	2	5	100

To obtain 50 per cent clay substance $\frac{50}{87}\times100=57.5$ per cent of china clay Cl have to be used; these contribute $\frac{57.5}{100}\times11=6.32$ per cent "ideal" feldspar and $\frac{57.5}{100}\times2=1.15$ per cent quartz. Therefore only $25.00-6.32=18.68$ per cent "ideal" feldspar is needed to make up the 25 per cent aimed at, thus $\frac{18.68}{95}\times5=19.63$ per cent of feldspar Fl. This provides $\frac{19.63}{100}\times5=0.98$ per cent quartz. We have already $0.98+1.15$ per cent (from the china clay)$=2.13$

per cent quartz so that only $25.00 - 2.13 = 22.87$ per cent quartz sand Ql have to be introduced.

The batch composition by percentage is thus:

China clay Cl	57.50
Feldspar Fl	19.63
Quartz sand Ql	22.87
	100.00

2. Glaze Calculations (See Chapter 7, p. 130).

2.1. Conversion of Weight Percentage Composition into Molecular Formula and Vice Versa

In order to be able to calculate the batch composition of a glaze from a given molecular formula it is necessary to know the molecular formula and molecular weights of the various ingredients to be used. Some of them are shown in the form of a molecular formula, e.g. calcium carbonate. It is convenient for ceramic calculations to use formulae which show the individual oxides, i.e. calcium carbonate is expressed as $CaO.CO_2$ instead of $CaCO_3$.

The silicate materials very rarely occur pure and the supplier does not usually give their molecular formulae but their ultimate chemical analyses. Let us consider a potash feldspar of the following percentage chemical analysis:

SiO_2	67.80
Al_2O_3	17.50
Fe_2O_3	0.05
CaO	0.30
K_2O	12.00
Na_2O	2.20
Loss	0.15
	100.00

To obtain the molecular formula we have to divide the percentage weights by their respective molecular weights, viz.

$$67.80 \; SiO_2 \quad : \quad 60.1 = 1.1261$$
$$17.50 \; Al_2O_3 \quad : \quad 102.2 = 0.1713$$
$$0.05 \; Fe_2O_3 \quad : \quad 159.7 = 0.000313$$
$$0.30 \; CaO \quad : \quad 56.1 = 0.00535$$
$$12.20 \; K_2O \quad : \quad 94.2 = 0.1275$$
$$2.20 \; Na_2O \quad : \quad 62.0 = 0.0355$$

As, according to the conventional presentation of molecular formulae, the sum of the "bases" is unity, the above molecular parts have to be divided by 0.16835. Written in the conventional form we thus have:

$$\left. \begin{array}{l} 0.032 \text{ CaO} \\ 0.757 \text{ K}_2\text{O} \\ 0.211 \text{ Na}_2\text{O} \end{array} \right\} 1.018 \text{ Al}_2\text{O}_3.6.690 \text{ SiO}_2.$$

Molecular weight 594.

The amounts of iron oxide in feldspars are usually so small that they can be neglected. Loss on ignition can also be disregarded.

Some potters prefer to judge glazes by their chemical weight percentage composition, and for this reason they may want to convert a given molecular formula into weight percentages. The molecular parts are multiplied by their respective molecular weights (their sum being the molecular weight of the glaze) and brought to percentage.

2.2. *Calculation of a Batch Composition Based on a Given Molecular Formula*

2.2.1. *Unfritted Glaze*

Let us take the molecular formula of glaze 2 on p. 132 (Chapter 7).

$$\left. \begin{array}{l} 0.7 \text{ CaO} \\ 0.3 \text{ K}_2\text{O} \end{array} \right\} \text{Al}_2\text{O}_3.10 \text{ SiO}_2$$

The feldspar at our disposal has the molecular formula of:

$$\left. \begin{array}{l} 0.032 \text{ CaO} \\ 0.757 \text{ K}_2\text{O} \\ 0.211 \text{ Na}_2\text{O} \end{array} \right\} 1.018 \text{ Al}_2\text{O}_3.6.680 \text{ SiO}_2$$

and the molecular weight of 594.

To calculate the batch composition of a glaze it is expedient to deal first with that oxide which is introduced by one material only. We thus start with 0.3 mol. parts of potash. We have to take the total alkalis in the feldspar available, viz. $0.757 + 0.211$ $Na_2O = 0.968$. The amount of potash introduced is $\dfrac{0.3}{0.968} \times 0.757 = 0.235$ mol. parts, and that of soda $\dfrac{0.3}{0.968} \times 0.211 = 0.065$ mol. parts, $0.235 + 0.065 = 0.3$ mol. parts total alkali. By analogous calculations we obtain the amounts of the other oxides in the feldspar, viz. 0.010 CaO, 0.316 Al_2O_3, and 2.07 SiO_2. The ratio $\dfrac{0.3}{0.968}$ equals 0.31 and can be used as a check for the sum of the bases, viz.

CaO	0.010
K₂O	0.235
Na₂O	0.065
	0.310

The amount of feldspar to be introduced is $0.31 \times 594 = 184.0$ weight parts. The feldspar contributes 0.01 mol. parts lime; we require 0.7 mol. parts, therefore 0.69 mol. parts are still needed. They are introduced by whiting ($CaO.CO_2$, mol. weight 100.1), thus $0.69 \times 100.1 = 69.0$ weight parts.

Dealing with alumina, the feldspar contributes 0.316 mol. parts and to make up the one molecular part required as shown in glaze 2, another 0.684 mol. parts have to be added. It is always desirable to have a certain percentage of china clay present in a glaze in order to keep it well in suspension. The amount should not exceed 10 per cent of the batch composition as otherwise the glaze becomes too plastic (which would lead to too slow drying and crawling). In porcelain glazes about 0.15 mol. parts of alumina are introduced as china clay ($Al_2O_3.2SiO_2.2H_2O$) in the raw form. The molecular weight of "ideal" china clay (clay substance) being 258.4, the amount of china clay to be introduced is $0.15 \times 258.4 = 38.8$ weight parts.

China clays invariably contain small amounts of alkali but with as low an amount as 0.15 mol. parts they, like ferric oxide and titania, can be neglected. The remaining $1 - (0.316 + 0.150) = 0.534$ parts of alumina required can be introduced by alumina (corundum) itself. The molecular weight of alumina (Al_2O_3) being 102.2, the amount of alumina to be introduced is $0.534 \times 102.2 = 55.5$ weight parts. (Very often calcined china clay—mol. wt. $258.4 - 36.0 = 222.4$—is used to make up the alumina content stipulated. If the alkalis in it exceed, say, 0.20 mol. parts, the amount of feldspathic material has to be reduced accordingly.)

Regarding silica, 2.07 mol. parts were contributed by the feldspar and 0.3 by the china clay; hence 7.63 mol. parts have to be added in the form of quartz (SiO_2, mol. wt. 60.1) corresponding to $7.63 \times 60.1 \simeq 460$ weight parts. (If calcined china clay instead of alumina had been added this would have brought in 1.068 mol. parts of silica so that 6.562 mol. parts silica would be required, thus $6.56 \times 60.1 \simeq 396$ weight parts quartz.)

The batch composition of the glaze is thus:

	Weight parts	%
Feldspar	184.0	22.78
Whiting	69.0	8.54
China clay (raw)	38.8	4.81
Calcined alumina	55.5	6.87
Quartz	460.0	57.00
	807.3	100.00

Of, if calcined china clay had been used:

	Weight parts	%
Feldspar	184.0	22.78
Whiting	69.0	8.54
China clay (raw)	38.8	4.81
China clay calcined	119.2	14.77
Quartz	396.0	49.10
	807.0	100.00

Because of the soda content in the feldspar used the mol. formula of the glaze does not correspond exactly to glaze 2 but is as follows:

$$\left. \begin{array}{l} 0.700 \text{ CaO} \\ 0.235 \text{ K}_2\text{O} \\ 0.065 \text{ Na}_2\text{O} \end{array} \right\} \text{Al}_2\text{O}_3.10 \text{ SiO}_2.$$

2.2.2. *Fritted Glaze*

Let us choose glaze 6, as shown on p. 133 (Chapter 7).

$$\left. \begin{array}{l} 0.3 \text{ PbO} \\ 0.1 \text{ MgO} \\ 0.4 \text{ CaO} \\ 0.1 \text{ K}_2\text{O} \\ 0.1 \text{ Na}_2\text{O} \end{array} \right\} \left. \begin{array}{l} 0.3 \text{ Al}_2\text{O}_3 \\ 0.4 \text{ B}_2\text{O}_3 \end{array} \right\} 3.0 \text{ SiO}_2$$

Firstly we have to make the frit and this is calculated exactly as if it were a glaze. We reduce the alumina content to 0.2 mol. parts to make allowance for a certain amount of china clay to be mixed with the finished frit to keep the actual glaze in suspension. Proceeding as before we get:

		Weight parts	%
0.3 PbO by lead oxide (PbO)	0.3×223.2	66.96	16.93
0.1 MgO by dolomite (MgO)			
⁃ CaO.2CO₂) also introducing 0.1 CaO	0.1×184.4	18.44	4.67
0.3 CaO by whiting (CaO.CO₂) (0.4−0.1)	0.3×100.1	30.03	7.67
0.1 K₂O by potassium nitrate (2KNO₃)	0.1×202.2	20.22	5.12
0.1 Na₂O by borax (Na₂O.2B₂O₃.10H₂O)			
also introducing 0.2 B₂O₃	0.1×381.8	38.18	9.66
0.2 B₂O₃ by crystalline boric acid (0.4−0.2)			
(B₂O₃.3H₂O)	0.2×123.6	24.72	6.35
0.2 Al₂O₃ by china clay raw			
(Al₂O₃.2SiO₂.2H₂O) also introducing			
0.4 SiO₂	0.2×258.4	51.68	13.08
2.4 SiO₂ by quartz (SiO₂) (2.8−0.4)	2.4×60.3	144.24	36.52
		394.47	100.00

During fritting considerable losses occur arising from water of crystallization, combined water as well as carbonate and nitrate, viz.

			Weight parts
Dolomite	2CO₂	: 0.2×88	17.6
Whiting	CO₂	: 0.3×44	13.2
Potassium nitrate	N₂O₅	: 0.1×108	10.8
Borax	10H₂O	: 0.1×180	18.0
Boric acid	3H₂O	: 0.2×54	10.8
China clay	2H₂O	: 0.2×36	7.2
Total loss			77.6

The molecular weight of the frit

$$\left. \begin{array}{l} 0.3\ PbO \\ 0.1\ MgO \\ 0.4\ CaO \\ 0.1\ K_2O \\ 0.1\ Na_2O \end{array} \right\} \left. \begin{array}{l} 0.2\ Al_2O_3 \\ 0.4\ B_2O_3 \end{array} \right\}\ 2.8\ SiO_2$$

is thus: $394.47 - 77.60 = 316.87$

To make the glaze itself we add:
0.1 Al_2O_3 by china clay raw: $0.1 \times 258.4 = 25.84$ weight parts.
The composition of the glaze is thus:

	Weight parts	‘%
Frit	316.87	92.5
China clay (raw)	25.84	7.5
	342.71	100.0

2.3. Calculation of Molecular Formula of a Given Batch Composition

The technologist may be asked to improve a glaze; only its recipe is known. Before he is able to start experiments he must know its molecular formula.

Let us consider the following batch composition by percentage (taking glaze 2 in reverse):

Potash feldspar	22.78
Whiting	8.54
China clay (raw)	4.81
Calcined alumina	6.87
Quartz	57.00
	100.00

The molecular formula of the potash feldspar is as follows:

$$\left. \begin{array}{l} 0.032\ CaO \\ 0.757\ K_2O \\ 0.211\ Na_2O \end{array} \right\}\ 1.018\ Al_2O_3.6.680\ SiO_2.$$

Molecular weight 594.

If we divide the percentage of potash feldspar by the molecular weight we obtain the total of the bases, viz. $\dfrac{22.78}{594} = 0.0384$ mol. parts. The individual molecular parts of the constituent oxides of the potash feldspar are thus:

CaO : $0.032 \times 0.0384 = 0.00123$ mol. parts
K_2O : $0.757 \times 0.0384 = 0.02907$ mol. parts
Na_2O : $0.211 \times 0.0384 = 0.00810$ mol. parts
Al_2O_3 : $1.018 \times 0.0384 = 0.0390$ mol. parts
SiO_2 : $6.680 \times 0.0384 = 0.2562$ mol. parts

Again we have a check by adding the molecular parts of lime, potash, and soda together, the sum of which is 0.0384, the total obtained in the first place.

With whiting (CaO.CO$_2$, mol. weight 100.1) we obtain $\frac{8.54}{100.1}=0.0854$ mol. parts of lime.

China clay (Al$_2$O$_3$.2SiO$_2$.2H$_2$O, mol. weight 258.4) adds $\frac{4.81}{258.4} \sim 0.0185$ mol. parts alumina and 0.0370 mol. parts silica.

Calcined alumina (Al$_2$O$_3$, mol. weight 102.2) contributes $\frac{6.87}{102.2}= 0.0671$ mol. parts Al$_2$O$_3$.

Quartz (SiO$_2$, mol. weight 60.1) introduces $\frac{57}{60.1} \sim 0.945$ mol. parts silica.

We thus have:

CaO   : 0.00123 mol. parts from feldspar
        0.08540 mol. parts from whiting   total 0.08663
K$_2$O   : 0.02907 mol. parts from feldspar   total 0.02907
Na$_2$O : 0.00810 mol. parts from feldspar   total 0.00810
Al$_2$O$_3$: 0.0390 mol. parts from feldspar
        0.0185   mol. parts from china clay
        0.0671   mol. parts from calcined alumina   total 0.12460
SiO$_2$: 0.2562 mol. parts from feldspar
        0.0370   mol. parts from china clay
        0.9450   mol. parts from quartz   total 1.23820

Converting the sum of the bases (0.1238 mol. parts) to unity we thus obtain (rounded off):

$$\left. \begin{array}{l} 0.70\ CaO \\ 0.24\ K_2O \\ 0.06\ Na_2O \end{array} \right\} \quad Al_2O_3 . 10\ SiO_2$$

Further Reading

Calculations in ceramics: **Griffiths and Radford (1965)**.

Glossary

Acid gold. A form of gold decoration; the glazed surface is etched with hydrogen fluoride (HF).

Ark. A large vat used for mixing or storing of clay slip.

Aventurine. A decorative effect on the surface of pottery by small bright coloured crystals, e.g. haematite.

Banding. The application of a line or band of colour to the edges of plates, etc.

Bat. (a) A re-shaped disc of body used in plate making. (b) A refractory slab for firing ware.

Bedder. A plaster-of-Paris shape for forming a bed of powdered alumina on which bone china plates are fired.

Biscuit (bisque). State of ware which has been fired but not glazed.

Bleb. A blister (bubble) on pottery.

Bloating. Increase in size of vitrified pottery due to overfiring.

Body. (a) A mixture of raw materials for shaping pottery. (b) The interior of pottery, as distinct from glaze.

Bossing. Part of ground-laying process, striking ware with a pad to remove brush marks.

Bottle oven. Bottle-shaped, old-fashioned coal-fired kiln.

Boxing. Placing cups rim to rim on one another to prevent distortion in firing.

Bright gold. Cheaper type of gold resinate for decoration.

Bristol glaze. Unfritted glaze for stoneware containing zinc.

Brongniart's formula. Formula to obtain solid content in a suspension.

Brushing. Removing of bedding material from ware after biscuit firing.

Bullers' ring. Heat work indicator in ring form (earthenware) whose shrinkage is determined.

Bung. A vertical stack of saggars or setters.

Burning. Alternative to firing (less suitable).

Burnish (or best) gold. Durable type of gold applied to glazed ware as suspension in oil, fired, and made bright by rubbing with agate, etc.

Casting. Shaping by pouring liquid body into a mould, the liquid solidifying subsequently.

Celadon. A glaze of characteristic green colour by iron oxide fired reducingly.

Chamotte. Synonymous with **grog** (q.v.).

Chert. Siliceous rock for pavers and runners of grinding pans.

China stone. Synonymous with **Cornish stone** (q.v.).

Chittering. Ruptures along edge or rim of ware.

Chrome-alumina pink. Pink colour consisting of alumina, zinc oxide, and chromia.

Chrome-tin pink. Pink colour consisting of tin oxide and chromia.

Contravec. An arrangement for blowing air into a tunnel kiln at exist end to counteract convection currents.

Coral red. Low-temperature colour consisting of lead chromate.

Cornish stone. Partly decomposed granite made up of quartz, feldspar, and fluorine minerals.

Crackle. Synonymous with **craquelé** (q.v.).

Crank. Refractory support for firing glazed flatware.

Craquelé. Fine hair cracks (crazing) in a glaze.

Crawling. Glaze defect, unglazed areas or lines.

Crazing. Fine surface cracks on glazes.

Cullet. Broken glass used in **parian** (q.v.).

De-airing. Removal of air from plastic body or casting slip.

Decalcomania. U.S. term for **lithography** (q.v.).

Dobbin. Turntable dryer for tableware.

Dottling. Placing of flatware for firing horizontally supported by refractory pins (thimbles).

Dunting. Crack formation in ware having been cooled too quickly after firing.

Engobe. Coating of slip applied to a body to improve its appearance.

Faience. Type of earthenware originally made in Faenza.

Fettling. Removal of seams before firing.

Flat. Short for flatware, e.g. plates, etc.

Flux. A substance bringing about melting reactions; added to bodies and glazes to reduce the firing temperature.

Frit. Main ingredient of lower temperature glazes, being a melt of constituents, quenched to form a glass.

Glost. Means "glazed".

Gottignies kiln. Multi-passage kiln invented by R. and L. Gottignies.

Grinding pan. Synonymous with **pan** (q.v.).

Grog. Fired and crushed refractory clay used in **saggars** (q.v.), other kiln furniture, and very plastic clay mixtures to reduce over-plasticity.

Ground laying. Decorating method giving uniform coating of colour by dusting powdered colour on area painted with oil.

Hard paste. Obsolete term for hard porcelain.

Heat work. Combined effect of temperature and time.

Hollow ware. Cups, tea pots, milk cans, etc.

Ironstone china. Vitrified body of high strength.

Jasper ware. Fine stoneware by Josiah Wedgwood containing up to 50 per cent barium sulphate.

Jet dryer. Ware dried by jets of warm air.

Jiggering. Making plates and other flat ware, the tool forming the outside (back) and the mould the inside (front).

Jollying. Making cups, sugar basins, etc., the tool forming the inside and the mould the outside.

Kiln furniture. Refractory shelving and smaller shapes of refractory material for the support of ware during firing.

Leatherhard. Partially dried ware when drying shrinkage has ceased.

Liquid gold. Synonymous with **bright gold** (q.v.).

Lithography. Decoration by transfers.

Low sol(ubility) **glaze.** Lead glaze in which not more than 5 per cent of lead oxide is soluble.

Manganese-alumina pink. Stain for bodies: calcined mixture of manganese carbonate, aluminium hydrate, and borax.

Mangle. Dryer, ware moving on trays vertically suspended between two endless chains.

Marl. Calcareous clay, also low-grade fireclay.

Maturing temperature. (a) Firing temperature at which body develops the desired degree of vitrification. (b) Firing temperature at which glaze constituents have formed a liquid which, on cooling, becomes a glass of the required surface brilliance.

Mazarin blue. Rich dark-blue colour for under- or on-glaze containing about 50 per cent cobalt oxide.

Muffle. Kiln consisting of refractory internal shell in which ware is placed thus protecting it from open flames.

Multi-passage kiln. A tunnel kiln with many adjacent passages through which ware travels in opposite directions.

Orton cones. Pyrometric cones (q.v.) used in the United States.

Pan (pan mill). Old type of grinding device consisting of a bottom formed by chert stones (pavers) and large blocks of chert (runners).

Parian. Porcelain consisting of up to 70 per cent feldspar and 30 per cent china clay and a small amount of **cullet** (q.v.) for casting figurines.

Paste. Hard or soft porcelain body (pâte).

Pâte-sur-pâte. Building up of a bas-relief by hand painting with slip.

Paver. See pan.

Peeling. Breaking away of glaze due to its too high compression.

Pin. Small refractory bar of triangular cross-section for placing in glazing fire.

Pin-hole. Burst bubble in glaze partially healed.

Pitchers. Broken pottery.

Plucking. Blemish where pointed supports have stuck to and removed glaze.

Pot bank. Pottery factory.

Pug. Machine for consolidating plastic body.

Pyrometric cone. A pyramid of ceramic mixture which under certain firing conditions bends so that the tip of the pyramid is level with the base.

Rational analysis. Mineralogical composition of pottery materials and bodies comprising clay substance, feldspar, and quartz.

Rearing. Glazed flatware set on edge during firing.

Réaumur porcelain. Porcelain made by devitrifying glass.

Refractory. Resistant to high temperatures.

Resist. A layer resistant to hydrofluoric acid, also a layer applied to areas to be left free from colour in ground laying and other decorations.

Roller-head machine. Shaping machine characterized by revolving shaping tool.

Rouge flambé. Red glaze produced by copper in reducing conditions.

Runner. See **pan.**

Saggar. Refractory container in which ware used to be (and still is being) placed for firing.

Saggar maker's bottom-knocker. Operative who beats out clay-grog mix for bottom of saggar.

Salt glaze. Glaze formed on stoneware towards end of firing by common salt being thrown into the fireboxes, volatilizing and reacting with body constituents.

Sang de bœuf. Similar to **rouge flamé** (q.v.).

Scouring. The cleaning and smoothing of surface of biscuit-fired ware in a revolving drum with coarse abrasive medium, e.g. **pitchers** (q.v.).

Seger cone. Pyrometric cone (q.v.) used in Europe.

Seger formula. Molecular formula applied to glazes consisting of RO (bases whose sum is 1), R_2O_3, and RO_2.

Setter. Type of **saggar** (q.v.) aimed at economizing in space, its underside conforming to the upper surface of the plate, etc., to be fired.

Sgraffito. Form of decorating, a pattern being produced by scratching through the engobe exposing differently coloured body.

Silver marking. Grey marks left by cutlery on glaze.

Slip. Aqueous suspension of pottery materials.

Slip trailing. Forming a pattern on ware by a viscous slip through a fine orifice.

Smalt. Blue pigment: fused mix of cobalt oxide, sand, and flux.

Snakeskin glaze. Decorative effect based on crawling obtained by glaze of very low expansion or high surface tension.

Soaking. Maintaining maximum firing temperature to obtain desired degree of chemical reaction.

Spit-out. Craters in glaze, an aggravated form of pinholes, usually on porous ware due to vapours in decorating fire.

Sponging. Removal of surface blemishes after making with moistened sponge.

Stamping. Method of decoration by rubber stamp.

Sticking-up. Joining together, e.g. handle to cup.

Sucking. Volatilization of lead oxide and other glaze constituents, the vapours being sucked into porous refractory.

Thimble. Conical refractory support with projection at base for **dottling** (q.v.) of plates.

Throwing. Shaping method, a ball of prepared body being thrown on to a revolving potter's wheel and worked into the desired shape.

Towing. Smoothing of outer edge of dried ware.

Trailing. See **slip trailing.**

Vitreous, vitrified. "Glassy", implying porosity of generally less than 0.5 per cent.

Wedging. Homogenizing clay or body by hand.

Whirler. (a) Flatware the centre of which has sagged during firing. (b) Small turntable for centring ware in decorating and for keeping slip agitated during casting.

Wreathing. Slightly raised crescent in slip-cast pieces.

Further Reading

Dictionary of ceramics: Dodd (1967).

References—Author Index

(Page numbers in *italics* refer to "Further Reading" lists)

Page

AB FORSHAMMER (1981), Data Sheets. — 30

AINSWORTH, L. (1956), A method for investigating the structure of glazes based on *219* surface measurement, *Trans. Brit. Ceram. Soc.* 55 (10), 661.

AITKEN, M. J. (1964), Remanent magnetism in ancient ceramics and pottery kilns, 5 *Proc. Brit. Ceram. Soc.* (2) *Magnetic Ceramics*, 143.

ALBERS-SCHOENBERG, E. (1960), Sources of modern ceramics, *Bull. Amer. Ceram.* 204 *Soc.* 39 (3), 136.

ALLEN, A. C. (1966), Kagaku Togyo widens scope with high alumina, *Ceram. Ind.* 87 47 (6), 62.

ALSTON, E. (1974), Stabilization and binding of glazes—a new approach, *Trans. J.* 71 *Brit. Ceram. Soc.* 73 (2), 51.

ALSTON, E. (1975), Dispersing agents suitable for the deflocculation of casting slip, 71 *Trans. J. Brit. Ceram. Soc.* 74 (8), 279.

ALT, P. (1972), The problems of decoration in the ceramic industry, *Interceram* 21 (4), *169* 302.

AMISON, A., and HOLMES, W. H. (1985), A review of drying methods and driers in 83, 84, the pottery industry, *Brit. Ceram. Trans. J.* 84 (5), 154. — 85, 86, *91*

ANON. (1966), Soviet Technology Bulletin, April 1966, *Ceramics* 17 (208), 48. 180

ANON. (1975), High temperature for gold by Heraeus, *Ber. Deut. Keram. Ges.* 52 (2), 168, *169* 43.

ANON. (1984), Hutschenreuther plans for the future, *Keram. Zeitschrift* 36 (10), 551. 221, 226

ANON. (1985), Report on ceramic week in Rimini, *Interceram* 34 (6), 41. 66

ANON. (1986), Gough introduces microwave casting machine, *Interceram* 35 (2), 54. 79

ANON. (1986a), *Health and Safety in Ceramics*, 2nd Edition, Institute of Ceramics, 221 Pergamon Press, Oxford.

ARLETT, R. H., DI VITA, S., and SMOKE, E. J. (1963), Ceramic materials and method 228 for making same, U.S. Patent 3.084.053.

ASTBURY, N. F. (1962), The problem of the clay column, *Claycraft* 36 (2), 40. 48

ASTBURY, N. F. (1966), [Some problems of the mechanics of ceramic materials, *Proc.* *219* *Brit. Ceram. Soc.* (6), 103.]

AUSTIN, L. K. (1976), Intermittent firing of bone china, *Trans. J. Brit. Ceram. Soc.* 114, *123* 77 (2), 45.

AYERS, J. (1964), *The Seligman Collection of Oriental Art*, 2, "Chinese and Korean *195* pottery and porcelain", Lund Humphries, U.S.A.

BAILEY, Sir R. (1986), Address at the Institute of Ceramics convention, Cambridge, 9 *228* April, 1986, *Brit. Ceram. Trans. J.* 85 (4), ix.

BARRELET, J. (1964), Porcelains de verre en France, *Cahiers de la Céramique* (36), 254. *195*

BARSBY, N. (1971), The "package" tunnel kiln, *Interceram* 20 (2), 142. 111

BARTRAM, C., and BRIGGS, D. (1982), A new horizon in process control with 228 microcomputers, *Trans. J. Brit. Ceram. Soc.* 81 (6), 171.

BASNETT, D. (1980), A new concept for offset printing, *Trans. J. Brit. Ceram. Soc.* 79 166, *169* (4), lvii.

BASNETT, D. (1987), The influence of particle orientation on slip casting, *Brit. Ceram.* 82 *Trans. J.* 86 (1), 9.

BATCHELOR, R. W. (1973), Flotation feldspar—the modern flux, *Trans. J. Brit.* 32 *Ceram. Soc.* 72 (1), 7.

BATCHELOR, R. W. (1974), Modern inorganic pigments, *Trans. J. Brit. Ceram. Soc.* 155, *169* 73 (8), 297.

BCRA (1967), British Ceramic Research Association, private communication.   197

BEAVER, S. H. (1964), The Potteries: a study in the evolution of a cultural lanscape,   228
The Institute of Brit. Geographers, Trans. & Papers, Publication 34.

BEECH, D. G. (1959), The constitution of bone china, *The A.T. Green Book*, (Brit.   37, 97,
Ceram. Res. Assoc.) 49.   123

BERG, P. W. (1963), The influence of kiln atmosphere during porcelain firing, *Ber.*   100, 188
Deut. Keram. Ges. **40** (7), 417.

BERNHARD, D. (1974), Correlation between lead release according to DIN 51 031 and   217, *219*
the amount of lead actually transferred into ordinary food, *Ber. Deut. Keram.*
Ges. **51** (6), 169.

BERNHARD, D. (1976), Glazes and decorating colours in the age of automatic   *219*
dishwashers, *Keram. Zeitschrift* **28** (5), 247.

BINNS, D. B. (1962), Some physical properties of two-phase crystal-glass solids,   *219*
Science of Ceram. **1**, 315.

BLANKENBURG, H. (1971), Comparison between pottery body preparation by   57
conventional methods and new preparation techniques employing spray-dried
granulate, *Interceram* **21** (2), 128.

BLASIUS, E., and others (1984), Investigations on the ageing of industrial casting slips,   71
CFI/Ber. Deut. Keram. Ges. **74** (8), 395.

BLOOR, W. A., and DINSDALE, A.(1962), Protective clothing as a factor in the dust   221, *228*
hazards of potters, *Brit. J. Ind. Med.* **19**, 229.

BOURRY, E. (1919), *A Treatise on Ceramic Industries*, translated by A. B. Searle, Scott   174
Greenwood & Son, London.

BRADSHAW, A., and GATER, R. (1980), Plastic forming in the tableware industry,   65, 67,
Ceram. Eng. Sci. Proc. **1** (9/10), 877.   82

BRETT, N. H., and others (1970), The thermal decomposition of hydrous layer   11
silicates and their related hydroxides, *Quart. Rev. Chem. Soc.* **24**, 155–207.

BRINDLEY, G. W., and NAKAHIRA, M. (1959), The kaoline–mullite reaction series, *J.*   10
Amer. Ceram. Soc. **42** (7), 319.

BRINDLEY, G. W., and others (1967), Kinetics and mechanism of dehydroxylation   11
processes. I. Temperature and vapor pressure dependence of dehydroxylation of
kaolinite, *Amer. Mineralogist* **52**, 201.

BUDWORTH, D. W. (1970), *An Introduction to Ceramic Science*, Pergamon Press,   37, *219*
Oxford, pp. 118–164 (Texture) and pp. 206–243 (Mechanical Properties).

BULL, A. C. (1982), Bodies, glazes and colours for fast firing, *Trans. J. Brit. Ceram.*   100, 167,
Soc. **81**, (3), 69.   168, *169*

CAMM, J. (1981), Particle size distribution and fine ceramics, *Interceram* **30** (5), 479.   57

CAMM, J., and WALTERS, W. L. (1983), Semi-vitreous china, fine earthenware,   180, *195*
Ceramic Monograph 2.1.5, *Handbook of Ceramics*, Verlag Schmid GmbH,
Freiburg i.Br., Federal Republic of Germany.

CASWELL, D. J. (1977), Advances in rapid tableware drying, *Interceram* **26** (2), 117.   86, *91*

CHARLES, R. (1964), *Continental Porcelain of the 18th Century*, Ernest Benn, London.   5, 106,
    195

CHATTERJEE, S. K. (1960), Birth of pristine pottery and the Hoary cultural   5
inheritance of India and Bengal, *Indian Ceramics* **7** (2), 37.

CHRONBERG, S. (1978), Electro-phoretic facilities for the fabrication of ceramic   78
products, *Ind. Ceram.* (718), 423.

CHRONBERG, M. S., and HÄNDLE, F. (1978), Processes and equipment for the   78, 79
production of materials by electro-phoresis 'Elephant', *Interceram* **27** (1), 33.

CIARRAPICO, J. O. (1979), Dry press 20 plates per minute automatically, *Ceram. Ind.*   79
113 (1), 36.

CLARK, N. O. (1964), Control of china clays, *J. Brit. Ceram. Soc.* **1** (2), 262.   16

CLARKE, F. J. P., TATTERSALL, H. G., and TAPPIN, G. (1966), Toughness of ceramics   119, *219*
and their work of fracture, *Proc. Brit. Ceram. Soc.* (6), 163.

CLOUGH, A. R. J., and SIMCOCK, H. (1976), The firing of clay-glazed cups, *Trans. J.*   122
Brit. Ceram. Soc. **75** (2), 36.

COBLE, R. L., and BURKE, J. E. (1963), Sintering in ceramics, *Progress in Ceramic*   123
Science (Pergamon Press) **3**, 197.

COPELAND, S. (1952), The high frequency drying of clayware, *Trans. Brit. Ceram.*   85
Soc. **51** (11), 573.

COPELAND, R. T. (1961), English porcelain, a comment and a reply, *Ceramics* **12**   *195*
(144), 44.

COUDAMY, J. (1977), A new generation of intermittent kilns, *Interceram* **26** (4), 270.   115, 119,
                                                                                             123

CRIADO, J. M., and others (1984), Re-examination of the kinetics of the thermal   11
dehydroxylation of kaolinite, *Clay Min.* **19** (4), 653.

CUBBON, R. C. P. (1982), Plastic body shaping processes and dust pressing in   65, 67,
tableware manufacture, *Trans. J. Brit. Ceram. Soc.* **81** (1), 9.   80, *82*

CUBBON, R. C. P. (1984), Tableware shaping processes, *Brit. Ceram. Trans. J.* **83** (5),   67, 77
121.

CUBBON, R. C. P., and TILL, J. R. (1980), Preparation of ceramic bodies, Ceramic   49–52,
Monograph 1.3, *Handbook of Ceramics*, Verlag Schmid GmbH, Freiburg i.Br.,   54, 57
Federal Republic of Germany.

CUBBON, R. C. P., and others (1981), The extraction of lead from ceramic tableware   217, *219*
by foodstuffs, *Trans. J. Brit. Ceram. Soc.* **80** (4), 125.

CUBBON, R. C. P., and others (1984), A quantitative test for dishwasher detergent   215, *219*
attack on tableware, *Interceram* **33** (1), 13.

DALE, A. J., and FRANCIS, M. (1943), The biscuit firing schedule for ceramic goods,   12
Trans. Brit. Ceram. Soc. **42** (3), 42.

DALE, A. J., and GERMAN, W. L. (1964), *Modern Ceramic Practice*, Scott Greenwood,   5
London.

DAMS, R. (1984), Fine grinding in mills, *Keram. Zeitschrift* **36** (12) 655.   57

DIETZEL, A. (1953), Bedeutung der Gleichgewichtsdiagramme für den Keramiker,   *219*
Sprechsaal **86**, 253.

DIETZEL, A., and HAAS, B. (1971), The influence of various oxides on the reactivity of   *149*
earthenware glazes with the body, *Ber. Deut. Keram. Ges.* **48** (12), 511.

DINSDALE, A. (1953), The development of firing in the pottery industry, *Ceramics, a*   *123*
Symposium, 363 (GREEN, A. T. and STEWART, G. H. (Eds.), Stoke-on-Trent).

DINSDALE, A. (1959), private communication.   75

DINSDALE, A. (1961), A note on the strength of bone china, *Bull. Amer. Ceram. Soc.*   197, 204
40 (1), 32.

DINSDALE, A. (1963), Crystalline silica in whiteware bodies, *Trans. Brit. Ceram. Soc.*   27
62 (4), 321.

DINSDALE, A. (1967), The constitution of bone china, *Science of Ceram.* **3**, 323. ·   *37, 98,*
                                                                                        123

DINSDALE, A. (1976), Translucency of tableware bodies, *Bull. Amer. Ceram. Soc.* **55**   211, 212,
(11), 993.   *219*

DINSDALE, A. (1981), Modern trends in whiteware processing, *Bull. Amer. Ceram.·*   228
Soc. **60** (2), 199.

DINSDALE, A. (1986), *Pottery Science, Materials, Processes and Products*, Ellis   *123*, 148,
Horwood Series, Applied Science and Industrial Technology.   *169, 219*

DINSDALE, A., and WILKINSON, W. T. (1966), Strength of whiteware bodies, *Proc.*   199, *219*
Brit. Ceram. Soc. (6), 119.

DINSDALE, A., CAMM, J., and WILKINSON, W. T. (1967), The mechanical strength of   200–202,
ceramic tableware, *Trans. Brit. Ceram. Soc.* **66** (8), 367.   *219*

DODD, A. E. (1953), The forms of silica, *Ceramics, a Symposium*, 201 (GREEN, A. T.   37
and STEWART, G. H. (Eds.), Stoke-on-Trent).

DODD, A. E. (1967), *Dictionary of Ceramics*, George Newnes, London.   102, 170,
                                                                        177–179

DORSCHNER, R., and STROBEL, K. (1970), Process-computer controlled central body   53
preparation plant at the porcelain factory of Lorenz Hutschenreuther A.G., Selb,
Ber. Deut. Keram. Ges. **47** (3), 176.

DUBE, A. (1980), A whiteware dream comes true: Isostatic pressing — a tool to   82
complete automation, *Ceram. Eng. Sc. Proc.* **1** (9/10), 882.

ELIAS, X., and POPPI, M. (1980), Operating cost comparison between conventional   120, *123*
and rapid firing tunnel kilns, *Interceram* **29** (1), 42.

ELLIS, K. A. (1973), Mechanized decorating within the ceramic industry, *Interceram*   *169*
    22 (4), 305.
ECC INTERNATIONAL LTD. (1984), Data Sheets.   14
EPPLER, R. A. (1977), Formulation and processing of ceramic glazes for low lead   216
    release, *Ind. Ceram.* (706), 362.
FERRARI, R. (1985), *Handbook for Ball Mill Grinding*, Faenza Editrice.   *57*
FISHLOCK, D. (1966), Towards tougher materials, *New Scientist* **3**, 283.   *228*
FLINDERS-PETRIE, SIR W. M. (1924–5), *Trans. Newcomen Soc.* **5**, 72.   *149*
FORD, R. W. (1964, 1986), Institute of Ceramics Textbook Series, 3. *Ceramic Drying*,   88, *91*
    2nd Edition, Pergamon Press, Oxford.
FORD, W. F. (1967), Institute of Ceramics Textbook Series, 4. *The Effect of Heat on*   *123*
    Ceramics, MacLaren & Sons, London.
FORRESTER, A. J. (1986), Impact of raw material changes on bone china manufacture,   *37, 195*
    Brit. Ceram. Trans. J. **85** (6), 180.
FRAHME, H. H. (1982), *Automatic Batching for Whitewares Production, Technical*   *228*
    Innovation in Whiteware, Alfred University Press Inc., p. 45.
FRANKLIN, C. E. L., and FORRESTER, A. J. (1975e, The development of bone china,   *195*
    Trans. J. Brit. Ceram. Soc. **74** (4), 141.
FREY, E., and SCHOLZE, H. (1979), Lead and cadmium release of fused colours,   217, *219*
    glazes and enamels in contact with acetic acid and foodstuffs and under the
    influence of light, *Ber. Deut. Keram. Ges.* **56** (10), 293.
GARDEIK, H. O., and SCHOLZ, R. (1981), Thermal technology for tunnel kiln firing in   *123*
    the ceramic industry. Ceramic Monograph 1.5.2, *Handbook of Ceramics*, Verlag
    Schmid GmbH, Freiburg i.Br. Federal Republic of Germany.
GAWLYTTA, M., and others (1981), Development of a high strength special gypsum   61
    for the making of moulds for pressing in the ceramic industry, *Silikattechnik*
    32 (12), 364.
GAY, P. W. (1978), Optimization in pottery manufacturing: a case study, *J. Opt. Res.*   *228*
    Soc. **29** (4), 323.
GERMAN, W. L. (1965), Properties of complex silicate polymers, *Ceramics* **16** (192),   *228*
    39.
GOONVEAN & ROSTORACK CHINA CLAY CO. LTD.(1984), Data Sheets.   14
GOULD, R. E. (1942), Making ceramic articles (by RAM pressing), U.S. Patent   66
    2,273,859. 24th February 1942.
GOULD, R. E. (1947), *Making True Porcelain Dinnerware*, Industrial Publications Inc.,   186
    Chicago.
GRAHN, T. (1983), Stoneware for domestic use, Ceramic Monograph 2.1.4,   *195*
    Handbook of Ceramics, Verlag Schmid GmbH, Freiburg i.Br., W. Germany.
GREEN, A. T. (1930), The contribution of Josiah Wedgwood to the technical side of   *228*
    the pottery industry, *Trans. Ceram. Soc.* **29** (5), 5.
GRIFFITHS, R., and RADFORD, C. (1965), *Calculations in Ceramics*, MacLaren,   237
    London.
GROENOU, A. BROESE VAN (1982), Theory of dust pressing, Part I: Models for die   82
    compaction, Ceramic Monograph 1.4.5.1.1., *Handbook of Ceramics*, Verlag
    Schmid GmbH, Freiburg, i.Br., W. Germany.
HAESE, U. (1985), Metal-free jet milling of ceramics and glasses in research and   42
    production, *Sprechsaal* **118** (6), 525.
HANDKE, H. (1974), The manufacture of pottery products on the injection-moulded   68, *82*
    principle, using permanent moulds, *Interceram* **23** (4), 260.
HARMS, W. (1976), The use of kiln furniture in rapid firing, *Ber. Deut. Keram. Ges.*   *123*
    53 (11), 391.
HARMS, W. (1979), Kiln furniture for fast firing processes, *Ber. Deut. Keram. Ges.* **56**   103, *123*
    (8), 223.
HARMS, W. (1984), Optimization of kilns for modern tableware production, *Keram.*   112, *123*
    Zeitschrift **36** (3), 123.
HARRIS, R. G. (1939), Observations on some causes and effects of vitrification in   203
    ceramic bodies, *Trans. Brit. Ceram. Soc.* **38** (6), 401.
HATHERSALL, L. G. (1984), Automatic holloware glazing, *Ceram. Ind. J.* **93** (1048),   142, *149*
    13.

HAUSCHILD, W., and others (1978), Colours for Fast Firing—their importance, *169*
possibilities and limits, *Intercerm* **27** (1), 45.
HAUTH, JNR., W. E. (1951), Crystal Chemistry in ceramics, *Bull. Amer. Ceram. Soc.* 7–9, 22,
30 (2), 47; (4), 137, 138, 140 and 141; (6) 203 and 204.   26
HC SPINKS CLAY CO. INC. (TENNESSEE, USA) (1981), Data Sheets.   20
HEBERLEIN, K. (1972), The electrostatic coating of glaze on porcelain and pottery, 143
Ber. Deut. Keram. Ges. **49** (2), 59.
HEBERLEIN, K. (1976), Experience with electrostatic glazing of tableware, *Ber. Deut.* 143, *149*
Keram. Ges. **53** (2), 51.
HELLOT, –. (1959), Rapport au Roi Louis XV, *Sprechsaal* **92**, 146.   110
HELSING, H. (1970), The liquidization of ceramic suspensions in the production of 52
bodies from sprayed granulate and slip, for turning and extruding, *Interceram* **19**
(3), 226.
HENNICKE, J., and HENNICKE, H. W. (1982), Electrophoresis, Ceramic Monograph *82*
1.4.3, *Handbook of Ceramics*, Verlag Schmid GmbH, Freiburg i.Br., Federal
Republic of Germany.
HERRMANN, K. R. (1964), Celsian porcelain, *Trans. IXth International Ceram.* 34, 98
Congress, 163, Brussels.
HERRMANN, K. R., and WICH, J. (1969), Ceramic slips from the industrial point of *82*
view, *Keram. Zeitschrift* **21** (9), 568.
HODGKINSON, H.R. (1962), The mechanics of extrusion, *Claycraft* **36** (2), 42.   48
HOGG, C. S. (1979), The application of Kubelka-Monk theory of the study of *219*
translucent systems, *Proc. Brit. Ceram. Soc.* **28**, 23.
HOHLWEIN, H. R. (1972), Electronic data processing in the ceramic industry, based 53, 57
on the application of a computer for body preparation in the pottery industry,
Interceram **21** (3), 189.
HOLDRIGE, D. A. (1953), The colloidal and rheological properties of clay, *Ceramics, a* 73
Symposium, 60 (GREEN, A. T. amd STEWART, G. H. (Eds.), Stoke-on-Trent).
HOLMES, W. H. (1969), Why rapid firing is possible, *J. Brit. Ceram. Soc.* **6** (2), 19. 118, *123*
HOLMES, W. H. (1973), The effect of fluorine on pottery bodies, *Trans. J. Brit.* 100
Ceram. Soc. **72** (1), 25.
HOLMES, W. H. (1978), Developments in pottery firing. *Trans. J. Brit. Ceram. Soc.* 119, *123*
77 (1), 25.
HOLMES, W. H. (1981), Kiln design for the future, *Proc. Silver Jubilee Conf. Inst. of* 119, *123*
Ceram. **4** (1), 54.
HOLMES, W. H., and others (1971), The B. Ceram. R. A. Quick-fire kiln, *Trans. J.* 119
Brit. Ceram. Soc. **70** (7), 237.
HÖLTJE, G. (1979), Unloading machine for kiln cars with free programmed control, 103
Interceram **28** (4), 415.
HOLMSTRÖM, N. G. (1980), Fast firing of triaxial porcelain, *Ceram. Eng. Sci. Proc.* **1** 101
(9/10), 780.
INZIGNERI, M., and PECO, G. (1964), Effect of mica content of kaolins on the 31, 97
vitrification of ceramic products, *Trans. IXth International Ceram. Congress*, 237,
Brussels.
JOHNSON, C. J. S. (1985), The needs of industry, *Brit. Ceram. Trans. J.* **84** (4), 115. *228*
JOHNSON, R. (1976), Some observations on flint v. silica sand in pottery bodies, *Trans.* 180
J. Brit. Ceram. Soc. **75** (1), 1.
JONES, J. A. (1979), The advantages and disadvantages of continuous and intermittent 113
firing of whiteware production, *Interceram* **28** (4), 403.
JUNG, R. (1983), Dry pressing of flatware, *Keram. Zeitschrift* **35** (8), 398.   79
KAETHER, H. U., and others (1983), New trends in kiln technology, *Interceram* **32** (5), 118, *123*
44.
KEELING, P. S. (1961), Geochemistry of the common clay minerals, *Trans. Brit.* 37
Ceram. Soc. **60** (9), 678.
KEELING, P. S. (1962), The common clay minerals as a conintuous series, *Science of* 37
Ceram. (London) **1**, 153.
KELSO, J. L., and THORLEY, J. P. (1941–43), The Potter's technique at Tell Beit 5
Mirsim, *The Annual of the American School of Oriental Research* **21** and **22**.

KENNY, J. B. (1954), *The Complete Book of Pottery Making*, Isaac Pitman & Sons, 58, 82
London.

KERHOF, F. (1983), Fundamentals of the mechanical strength and the fracture 219
mechanics of ceramics, Ceramic Monograph 3.1.1, *Handbook of Ceramics*, Verlag
Schmid GmbH, Freiburg i.Br., Federal Republic of Germany.

KERSTAN, W. (1982), Reactions between body and glaze, *Keram. Zeitschrift* 34 (10), 149
584.

KIEFFER, R. (1979), Shaping by plastic pressing of ceramic bodies in an oil hydraulic 67
stamping press, *Keram. Zeitschrift*, 31 (3), 156.

KINGERY, W. D. (1959), Sintering in the presence of a liquid phase, *Kinetics of High* 123
Temperature Processes, 187, Technology Press M.I.T. and John Wiley, New
York.

KINGERY, W. D. (1960), Sintering in the presence of a liquid phase, *ibid.*, 131.   50, *123*

KINGERY, W. D., NIKI, E., and NARASIMKAN, M. D. (1961), Sintering of oxide and 123
carbide metal compositions in the presence of a liquid phase, *J. Amer. Ceram.
Soc.* 44 (1), 29.

KINGSBURY, P. C. (1939), Stoneware in the Electrochemical field, *Trans. Electrochem.* 197
Soc. 17, 131.

KLEIN, U., and WILKE, H. G. (1979), Fast firing—a challenge to the manufacturers 103
of kiln furniture, *Ber. Deut. Keram. Ges.* 56 (8), 220.

KNOTT, P. (1983), The glassy state, Ceramic Monograph 3.3.1.3, *Handbook of* 149
Ceramics, Verlag Schmid GmbH, Freiburg I.Br., Federal Republic of Germany.

KONTA, J. (1979), Deposits of ceramic raw materials, Ceramic Monograph 1.1.3, *37*
Handbook of Ceramics, Verlag Schmid GmbH, Freiburg i.Br., Federal Republic
of Germany.

KONTA, J. (1980), Properties of ceramic raw materials, Ceramic Monograph 1.1.4, *37*
Handbook of Ceramics, Verlag Schmid GmbH, Freiburg i.Br., Federal Republic
of Germany.

KOENIG, J. H. (1952), Comparison of some properties of plastic and china tableware, 227
Ceram. Age 59 (4), 15.

KRAJEWSKI, A., and RAVAGLIOLI, A. (1980), The influence of the ceramic body on *219*
lead release from decorated glaze, *Ber. Deut, Keram. Ges.* 57 (4/5), 76.

KURE, F. (1957), Glaze defects, *Interceram* (6), 34.   *149*

LAMBERT, M. (1974), Without computer or programmer, electrostatic glazing is now 143, *149*
at the service of industry for the glazing of about thirty different parts, *Interceram*
23 (2), 107.

LEVIN, E. M., ROBBINS, C. R., and MCMURDIE, H. F. (1964), *Phase Diagrams for* 123
Ceramists, Amer. Ceram. Soc., Columbus.

LITZOW, K. (1982), History of ceramic technology, Ceramic Monograph 5.1, 5, 71
Handbook of Ceramics, Verlag Schmid GmbH, Freiburg, i.Br, Federal Republic
of Germany.

LIVERANI, G. (1959), Premières porcelains européennes; Les essais des Médicis, *195*
Cahiers de la Céramique 15, 141.

LOHMEYER, S. (1968), Measuring damage to decoration on porcelain surfaces, *Ber.* 215, *219*
Deut. Keram. Ges. 45, 25.

LOHMEYER, S. (1969), Glazes and decorating colours in the age of automatic dish- 215, 219
washers, *Keram. Zeitschrift* 28 (5), 247.

LOVATT, J. W. (1984), Refractories for fast firing kilns, *Ceram. Ind. J.* 93 (1045), 21. 105, *123*

LUCHS, D. (1985a), Pressure casting—a new dimension, *Keram. Zeitschrift* 37 (4), 77
187.

LUCHS, D. (1985b), An interesting ceramic week in Italy, Faenza and Rimini, 166, *169*
Interceram 34 (6), 40 and 41.

LUNDIN, S. T. (1959), *Studies in Triaxial Whiteware Bodies*, 189, Almquist and *123*
Wiskell, Stockholm.

LYNG, S., and GAMLEM, K. (1974), Anorthosite as a ceramic material. I. Extrudable 31
and castable electrical porcelain and vitreous china. *Trans. J. Brit. Ceram. Soc.*
73 (5), 133.

MACKENZIE, K. J. D., and others (1985), Outstanding problems in the kaolinite- 11
mullite reaction sequence: Pt.1. Metakaolinite, *J. Amer. Ceram. Soc.* 68 (6), 293.

MAGET, H. (1979), Saggar filling and drawing equipment especially for unfired and   103
fired tableware including kiln furniture, *Ber. Deut. Keram. Ges.* **56** (8), 215.

MASSON, R. (1948), Aging of thermally stressed, dense ceramic bodies, *Schweiz.*   207
Mineralog. Petrog. Mitt. **28** (1), 303.

MASSON, R. (1956), Gefügespannungen und Zugfestigkeit von Hartporzellan, *Trans.*   *219*
Vth International Ceram. Congress, 347, Vienna.

MASSON, R. (1962), Contribution to the question of microstress in porcelain bodies,   *219*
Trans. VIIIth International Ceram. Congress, 393, Copenhagen.

MATSON, F. R. (1965). *Ceramics and Man*, Aldine Publishing, Chicago.   5

MATTHEWS, A. G. (1977), Productivity in once-firing tableware, *Brit. Ceram. Rev.*   228
(30), 16.

MAYER, F. K., and KAESTNER, F. (1930), Sprechsaal, **63**, 118.   7

MCMILLAN, P. W. (1964), *Glass-Ceramics*, Academic Press, London.   194

MELLOR, J. W. (1935), The durability of pottery frits, glazes, glasses and enamels in   13, *149*
service, *Trans. Brit. Ceram. Soc.* **34** (2), 113.

METTKE, P. (1984), The colouring of glazes, *Keram. Zeitschrift* **36** (10), 538.   169

METZEL, H. (1962), The firing of porcelain in tunnel kilns fired by liquid gas, *Ber.*   123
Deut. Keram. Ges. **39** (12), 589.

MIELDS, M. (1965), Gedanken zur Porzellanerfindung in Europa, *Sprechsaal* **98** (7),   183, *195*
147.

MITCHELL, D. (1974), The changing pattern of raw materials supplies—ball clay,   37
Trans. J. Brit. Ceram. Soc. **73** (8), 287.

MOERTEL, H. (1977), Porcelain for fast firing, *Ceramurgia International*, **3** (2), 65.   101

MOLNAR, E., and WAGNER, E. (1974), Abradability of porcelain glazes as a function   *219*
of grain size distribution and firing temperature, *Trans. XIIIth Intern. Ceram.*
Congress, Amsterdam.

MONTEROS, J. E. DE LOS, and others (1978), Siliceous porcelains without quartz, *Bol.*   186
Soc. Esp. Ceram. Vidrio **18** (3), 143.

MOORE, F. (1965), *Institute of Ceramics Textbook Series*, 2. *Rheology of Ceramic*   82
Systems, 41, MacLaren & Sons, London.

MOORE, H. (1956), Structure and properties of glazes, *Trans. Brit. Ceram. Soc.* **55**   *149*
(10), 589.

MOREY, G. W. (1954), *The Properties of Glass*, Reinhold Publish. Corp., New York.   *149*

MURPHY, J. M. (1983), Tableware producers reap benefits of tile's disk glazing,   142, *149*
Ceram. Ind. **121** (5), 33.

NASSETTI-USMAC (1985), Granulation saves 84% of energy, *Interceram* **34** (6), 39.   57

NETSCH, A. (1978), Reduction in heat consumption through use of jet dryers in the   86, *91*
fine ceramics industry, *Interceram* **27** (1), 31.

NIFFKA, H. (1971), First automatic plate production with fully automatic production   66
line, *Interceram* **20** (3), 194.

NIFFKA, H. (1978). A revolution in the manufacture of tableware. Successful isostatic   82
dry pressing of flatware, *Interceram* **27** (3), 185.

NIFFKA, H. (1979), New technologies in the manufacture of dinner-ware, *Keram.*   82
Zeitschrift **31** (7), 392.

NILSSON, H. O. (1976), Dry body preparation in the pottery industry—a new concept   50
in European whiteware production, *Trans. J. Brit. Ceram. Soc.* **75** (6), 143.

NOBLE, F. R., and others (1979), Studies in the mineralogy of ceramic clays, *Proc.*   16
Brit. Ceram. Soc. **28**, 117.

NORFLOAT, A/S (1981), Data Sheets.   30

NORRIS, A. W. (1981), The technological future of the whiteware industries, *Silver*   226, *228*
Jubilee Ed. J. Inst. Ceram. **4** (1), 30.

NORSK NEFELINE (1985), Data Sheets.   30

NORTON, F. H. (1952), *Elements of Ceramics*, Addison-Wesley Press, Cambridge,   5
Mass., U.S.A.

NORTON, F. H. (1958), Ceramics and the future, *Trans. Brit. Ceram. Soc.* **57** (8), 463.   *228*

OBERLIES, F., and POHLMANN, G. (1958), On the effect of micro-organisms on   47
feldspars, kaolin and clay, *Trans. VIth International Ceram. Congress*, 149,
Wiesbaden.

ODELBERG, A. S. W. (1931), Note on the durability of bone china hotel ware, *Trans.*   204
Ceram. Soc. **30** (6), 225.

OLGUN, C., and others (1970), Investigations on the use of pressure casting in the   76, *82*
porcelain industry, *Ber. Deut. Keram. Ges.* **47** (2), 105.

OTT, R. (1975), The RAM process, *Interceram* 24 (3), 219.   66, *82*

PAPEN, E. L. J. (1967), Isostatic pressing, *Ber. Deut. Keram. Ges.* **44** (3), 82.   82

PARMELEE, C. W. (1951), *Ceramic Glazes*, International Publications Inc., Chicago.   *149*

PFAFF, P. (1973), High temperature rapid firing for porcelain, *Interceram* 22 (2), 106.   168, *169*

PFAFF, P. (1985), Recent tendencies in fast firing decoration of porcelain, *Keram.Z.* 37   167, 168,
(2), 63.   *169*

PFUHL, H. (1976), Modern shaping systems for flat-ware, saucers and plate   66
production plants, *Interceram* 25 (3), 173.

PFUHL, H. (1977), Spraying plant for the glazing of porcelain flatware in the biscuit-   142, *149*
fired or white-dried state, *Keram.Z.* 29 (7), 343.

PHELPS, G. W. (1982), Slip casting, Ceramic Monograph 1.4.2, *Handbook of*   *82*
Ceramics, Verlag Schmid GmbH, Freiburg i.Br, Federal Republic of Germany.

POLO, M., and LATHAM, R. (1958), *The Travels of Marco Polo*, 210, Penguin,   *195*
London.

QUON, D. H. H., and BELL, K. E. (1980), Release of lead from typical Canadian   216
pottery glaze formulations, CANMET Rep. 80–7E.

RADFORD, C. (1961), English porcelain, a comment and a reply, *Ceramics* 12 (144), 44.   *195*

RADFORD, C., and KEATING, J. R. (1985), The use of coarse micaceous china clay   31
fractions as fluxing material in whiteware bodies, *Interceram* 34 (6), 31.

RADO, P. (1964), The evolution of porcelain, *J. Brit. Ceram. Soc.* **1** (3), 417.   181, *195,*
    228

RADO, P. (1970), Man and pottery, *J. Brit. Ceram. Soc.* 7 (4), 111.   *228*

RADO, P. (1971), The strange case of hard porcelain, *Trans. J. Brit. Ceram. Soc.* **70**   *195*
(4), 131.

RADO, P. (1981), Bone china, Ceramic Monograph 2.1.3, *Handbook of Ceramics*,   *195*
Verlag Schmid GmbH, Freiburg i.Br, Federal Republic of Germany.

RADO, P. (1986), The durability of pottery, *Interceram* 35 (2), 48.   *219*

RADO, P. (1987), The effect of detergents on porcelain, *Interceram* 36 (1), 24.   *219*

RAVAGLIOLI, A., and VECCHI, G. (1981), Majolica pottery, Ceramic Monograph 2.1.6,   *195*
Handbook of Ceramics, Verlag Schmid GmbH, Freiburg i.Br, Federal Republic
of Germany.

REH, H. H. (1966), Vitreous china tableware bodies in literature and practice,   *195*
Sprechsaal **99** (18), 784 and (21), 975.

REICHL, H. (1979), Automatic saggar filling and drawing equipment especially for the   103
handling of plates, *Ber. Deut. Keram. Ges.* **56** (8), 217.

REMMER, F. (1980), Production of quality crockery by the single firing process,   122, *123*
Keram. Zeitschrift 32 (10), 584.

RHODES, D. (1971), *Kilns—Design, Construction and Operation*, Pitman Publishing,   *123, 195*
London.

RHODES, D. (1973), *Clay and Glazes for the Potter*, Chiltern Book Co., Philadelphia.   *5, 195*

RIES, H. B. (1973), Preparation of fine ceramic bodies in the counter-current   49, 57
intensive mixer with high-energy rotor, *Interceram* 22 (3), 207.

RIES, H. B. (1979), Body preparation in fine ceramics, *Keram. Zeitschrift* 31 (7), 386.   50

RIGBY, G. R. (1949), The application of crystal chemistry to ceramic materials, *Trans.*   23, 24,
Brit. Ceram. Soc. 48 (1), 1.   98

ROBERTS, W. (1965), The micro-indentation hardness of glazes, *Trans. Brit. Ceram.*   *219*
Soc. **64** (1), 33.

ROBERTS, W. (1974), Dry glazing using a fluidized bed, *Trans. J. Brit. Ceram. Soc.* 73   143, 144,
(2), 47.   *149*

ROBERTS, W. (1981/4) and (1981/5), The development of automated off-set printing   164, 165,
for tableware, *Interceram* 30 (4), 400 and (5), 498.   *169*

ROBERTS, W., and MARSHALL, K. (1970), Glaze-body reactions: electrochemical   129, *149*
studies, *Trans. Brit. Ceram. Soc.* **69** (6), 221.

ROBERTS, W., and HOLMES, W. H. (1982), Keeping tableware within the limits,   216, 219
Ceram. Ind. J. **91** (1031), 10.

ROSENTHAL, E. (1949), *Pottery and Ceramics*, Penguin, London. 5, 195

ROYSTON, M. G., and BARRETT, L. R. (1958), Some observations and reservations on 207, 219 thermal shock theory, *Trans. Brit. Ceram. Soc.* **57** (10), 678.

RYAN, W. (1978), *Properties of Ceramic Raw Materials*, 2nd Edition, LICET, 37 Pergamon Press, Oxford.

RYAN, W. (1981), Fabrication by electrophoresis, *Trans. J. Brit. Ceram. Soc.* **80** (2), 77–79, 46. 82

ST. PIERRE, P. D. S. (1955), Reactions in bone china bodies, *J. Amer. Ceram. Soc.* **38** 98 (6), 217.

ST. PIERRE, P. D. S. (1954), Constitution of bone china, I. High temperature phase 98 equilibrium studies in the system tricalcium phosphate-alumina-silica, *J. Amer. Ceram. Soc.* **37**, 243.

ST. PIERRE, P. D. S. (1956), Reactions in bone china bodies—Discussion, *Trans. Brit.* 98 *Ceram. Soc.* **55**, 13.

SALMANG, H. (1961), *Ceramics, Physical and Chemical Fundamentals*, translated by 5 FRANCIS, M., Butterworths, London.

SAVAGE, G. (1959), *Pottery Through The Ages*, Penguin, London. 5, 195

SAVAGE, G. (1963), *Porcelain Through The Ages*, Penguin, London. 195

SCHACKNIES, G. (1983), The use of microprocessors, *Interceram* **32** (2), 28. 228

SCHLEGEL, W. (1975), Tendencies in the development of the technology of the 80 porcelain and earthenware plate—problems and criticism, *Sprechsaal* **108** (718), 220.

SCHEREZ, E. (1979), Factors of influence when employing transfer pictures with on- 169 glaze decorations, *Interceram* **28** (4), 412.

SCHNETTLER, F. J., MONFORTE, F. R., and RHODES, F. F. (1967), Ceramic raw 228 materials prepared for manufacturing processes by quick freeze method, *Chem. Processing* **13** (9), 18.

SCHOICK, E. C. VAN (1963), *Ceramic Glossary*, Amer. Ceram. Soc., Columbus. 178

SCHOLZ, R., and GARDEIK, H. O. (1980), Drying processes for porous materials, 91 Ceramic Monograph 1.5.1, *Handbook of Ceramics*, Verlag Schmid GmbH, Freiburg i.Br, Federal Republic of Germany.

SCHÜLLER, K. (1962), The influence of quartz on the stresses of porcelain, *Ber. Deut.* 219 *Keram. Ges.* **39** (5), 286.

SCHÜLLER, K. (1964), Reactions between mullite and glassy phase in porcelains, 95, 123 *Trans. Brit. Ceram. soc.* **63** (2), 103.

SCHÜLLER, K. (1967), Porcelain bodies of high strength on the basis of quartz or 219 cristobalite, Part 3, *Ber. Deut. Keram. Ges.* **44** (8), 387.

SCHÜLLER, K. H. (1976), Technological classification of kaolins according to their 13 mineral composition, *Ber. Deut. Keram. Ges.* **53** (10), 285.

SCHÜLLER, K. H. (1978), Crystallization of secondary mullite during firing of kaolins 123 of different purities, *Ber. Deut. Keram. Ges.* **55** (2), 52.

SCHÜLLER, K. H. (1979), Porcelain, Ceramic Monograph 2.1.1, *Handbook of* 123, 195 *Ceramics*, Verlag Schmid GmbH, Freiburg i.Br, Federal Republic of Germany.

SCHÜLLER, K. H., and others (1977), Scope and problems of rapid firing of colours on 169 porcelain, *Ber. Deut. Keram. Ges.* **54** (4), 101.

SCHÜLLER, K. H., and JÄGER, H. (1979), The chemistry and properties of feldspars 97 and their effect on porcelains, *Ber. Deut. Keram. Ges.* **56** (2), 29.

SCHÜLLER, K., and LINDL, P. (1964), A study of thermal expansion of ceramic 219 materials, Trans. IXth International Ceram. Congress, Brussels, 139.

SEGER, H. A. (1902), *Collected Writings*, 2, 577, The Chemical Publishing Co., 5 Easton, Pa.

SHAW, K. (1966), Colour chemists get down to fundamentals, *Ceramics* **17** (207), 22; 169 *ibid.* (208), 36.

SHAW, K. (1968), *Ceramic Colours and Pottery Decoration*, MacLaren & Sons, 169 London.

SHEPARD, A. O. (1956), Ceramics for the archaeologist, *Publication* 609, *Carnegie* 5 *Institution of Washington*, 360, Washington, D.C.

SINGER, F. (1910), Concerning the position of boron in the glaze formula, *Trans.* 130 *Amer. Ceram. Soc.* **12** (February).

SINGER, F. (1936), Un des problèmes de l'industrie de la vaisselle de porcelaine, *La* 203
Céramique (February).
SINGER, F. (1951), Chemical stoneware to-day, *Ceramic Age* **58** (6), 33.   176
SINGER, F. (1954), Low temperature glazes, *Trans. Brit. Ceram. Soc.* **53** (7), 396.   *149*
SINGER, F., and GERMAN, W. L. (1960), *Ceramic Glazes*, Borax Consolidated Ltd., *149*
London.
SINGER, F., and SINGER, S. S. (1963), *Industrial Ceramics*, Chapman & Hall, London. *5, 37,*
57, 82,
91, 111,
123, 149,
152, *169,*
183, *195*
SKARBYE, H. (1964), Strength of high tension porcelain, *Trans. IXth International* 219
Ceram. Congress, 149, Brussels.
SKARBYE, H. (1966), Influence of glaze on strength of high tension porcelain, *Trans.* 202, *219*
Xth International Ceram. Congress, 167, Stockholm.
SKRILETZ, R. A. (1977), New development in ram pressing dies, *Proc. 1977 Am.* 67
Ceram. Soc. Ann. Meeting and *1977 Joint Fall Meeting*, Material and Equipment
and Whitewares Division, Edited by W. C. Mohr, Columbus, Ohio, ACS, 1977,
p. 13.
SLADEK, R. (1986), The fast firing process in the ceramic industry, *Interceram* **35** (4), *123*
35.
SLAWSON, R. J. (1948), private communication.   74
SLINN, G., and RODGERS, K. (1980), Grinding of glazes, *Interceram* **29** (3), 398.   139, *149*
SMITH, A. N. (1953), The structure of glazes, *Ceramics, A Symposium*, 284 (GREEN, *149*
A. T. and STEWART, G. H. (Eds.), Stoke-on-Trent).
SOLON, L. (1907–8), Porcelain, *Trans. English Ceram. Soc.* **7**, 153.   179, 189,
195
SOSMAN, R. B. (1955), New and old phases of silica, *Trans. Brit. Ceram. Soc.* **54** (11), 37
655.
SPEER, M. (1981), Rapid firing plants, *Interceram* **30** (2), 106.   112, *123*
STRADLING, G. N., and others (1972), The effect of foodstuffs and related compounds 217
on the durability of glazes containing natural thorium, natural or depleted
uranium, *Trans. J. Brit. Ceram. Soc.* **71** (6), 171.
TAIT, H. (1962), *Porcelain*, 42, Spring Art Books, Paul Hamlyn, London.   *195*
TAKASHIMA, H. (1985), Relation between glaze structure and lead pollution, 216, *219*
Interceram **34** (1), 15.
TAKEDA, R., and others (1974), Glazes for low thermal expansion lithia ceramics and 127
new apparatus for glazing, *Rep. Res. Lab. Asahi Glass Co. Ltd.* **24** (1), 45.
TAYLOR, D., and others (1979), Replacing bone ash in bone china, *Trans. J. Brit.* 193
Ceram. Soc. **78** (5), 108.
TAYLOR, J. R. (1967), Colour in glazes and glasses, *J. Brit. Ceram. Soc.* **4** (2), 201. 169
TAYLOR, J. R., and BULL, A. C. (1986), Institute of Ceramics Textbook Series, *149*
Ceramic Glaze Technology, Pergamon Press, Oxford.
THOMAS, J. (1933), Josiah Spode—his times and triumphs, *Trans. Brit. Ceram. Soc.* 195, *228*
32 (11), 489.
THORMANN, P. (1982), Basic principles for the planning of the lay-out of ceramic *228*
factories, Ceramic Monograph 5.2, *Handbook of Ceramics*, Verlag Schmid
GmbH, Freiburg i.Br, Federal Republic of Germany.
THROCKMORTON, P. (1962), Oldest known shipwreck yields bronze age cargo, 2
National Geographic, **121** (5), 698.
TOWNSEND, M., and LUKE, K. (1985), Halloysite—a refined clay from New Zealand, 13
Ceram. Ind. J. **94** (1052), 29.
TRAWINSKI, H. (1979), Wet preparation of kaolin (china clay), Ceramic Monograph 15, 16,
1.2.1, *Handbook of Ceramics*, Verlag Schmid GmbH, Freiburg i.Br. Federal 37
Republic of Germany.
TRAWINSKI, H. (1981), The refining of kaolin, part II, Ceramic Monograph 1.2.1, *37*
Handbook of Ceramics, Verlag Schmid GmbH, Freiburg i.Br, Federal Republic
of Germany.

VAUGHAN, F., and DINSDALE, A. (1962), Moisture expansion, *Trans. Brit. Ceram.*   *219*
Soc. **61** (1), 1.

VIVIAN, H. E. (1980), A new approach to improving the grinding efficiency of ball   42, 80
mills, *Interceram* **29** (3), 386.

WADE, T. B., and BORTY, S. A. (1966), Effect of geometry on mechanical properties,   *219*
Bull. Amer. Ceram. Soc. **45** (8), 753.

WALKER, E. G., and DINSDALE, A. (1959), The physical basis of the casting process,   *74, 75*
The A.T. Green Book (Brit. Ceram. Res. Assoc.), 142.

WAYE, B. E. (1973), Automatic cup making, *Interceram* **22** (4), 271.   66, 86,
    91

WEAVER, K. F. (1967), Magnetic clues help date the past, *National Geographic* **131**   *5*
(5), 696.

WEBSTER, A. V., and others (1987), The properties of milled bone, *Brit. Ceram.*   *37*
Trans. J. **86** (3), 91.

WEDGWOOD, SIR, J. (1964), Josiah Wedgwood and his times, *Ber. Deut. Keram. Ges.*   *228*
41 (10), 592.

WEISS, R. (1979), The raw material quartz and its preparation, Ceramic Monograph   27, *37*
1.2.4, *Handbook of Ceramics*, Verlag Schmid GmbH, Freiburg i.Br, Federal
Republic of Germany.

WELLS, A. F., *Structural Inorganic Chemistry*, Clarendon Press, Oxford.   27

WEST, R., and GEROW, J. V. (1971), Estimation and optimization of glaze properties,   *149*
Trans. J. Brit. Ceram. Soc. **70** (7), 265.

WEYL, W. A. (1941), Phosphate in ceramic ware, II. Role of phosphates in bone   209
china, *J. Amer. Ceram. Soc.* **24**, 245.

WHITE, J. (1965), Sintering—an assessment, *Proc. Brit. Ceram. Soc.* **3**, 155.   *123*

WHITE, R. P. (1953), Saggars and kiln furniture, *Ceramics, A Symposium*, 303   102, *123*
(GREEN, A. T. and STEWART, G. H. (Eds.), Stoke-on-Trent).

WHITMORE, M. (1974), Spraying earthenware flatware, *Trans. J. Brit. Ceram. Soc.* **73**   142, *149*
(4), 125.

WIEDEMANN, T. (1966), Highly resistant porcelains, *Sprechsaal* **99** (11), 428.   *219*

WIEDEMANN, T. (1967, 1968, 1969), Pottery bodies, *Sprechsaal* **100**, 6 and 15; **101**, 11   *195*
and 15; **102**, 13 and 19.

WILLIAMSON, W. O. (1980), Feldspathic of feldspathoidal fluxes and their   28, 32,
beneficiation, Ceramic Monograph 1.2.3, *Handbook of Ceramics*, Verlag Schmid   *37*
GmbH, Freiburg i.Br, Federal Republic of Germany.

WILSON, S. T. (1916–17), The firing of pottery ovens, *Trans. Ceram. Soc.* **16**, 304.   192

WOLF, J. (1937), Basic materials used as colouring agents for clayware, etc.,   *169*
Sprechsaal **70** (48, 49, 50), 601, 612, 625.

WOOD, D. J. (1985), New developments for the ceramic industry from ECC   16, 55,
International, *Ceram. Ind. J.* **94** (1052), 34.   57

WOOLLEY, SIR, L. (1937), *Digging Up the Past*, Penguin, Harmondsworth.   *5*

WORCESTER ROYAL PORCELAIN CO. LTD., and LUNN, T. W. (1960), A new or   157
improved method of and apparatus for the manufacture of coloured ceramic
ware, *Brit. Patent* 831, 330.

WORRALL, W. E. (1982), Institute of Ceramics Textbook Series: 1. *Ceramic Raw*   37, 72
Materials, 2nd Edition, Pergamon Press, Oxford.

WORRALL, W. E. (1986), *Clays and Ceramic Raw Materials*, 2nd Edition, Elseveir   22, 25,
Applied Science Publishers, London.   37, 151

WRIGHT, G. E. (1960), The last thousand years before Christ, *National Geographic*   2
118 (6), 833.

YULE, SIR H. (1921), *The Book of Ser Marco Polo*, 2, Book Second, 66, John Murray,   *195*
London.

ZACHARIASEN, H. (1932), The atomic arrangement in glass, *J. Amer. Chem. Soc.* **54**,   124
3841.

ZEIDLER, H. (and SELTMANN, H.) (1972), Method and apparatus for making and   68
attaching handles for cups, pots and the like (injection moulding ceramic
handles), Brit. Pat. 1,274,468 (17/5/72).

ZIMMERMANN, H. (1987), Dry pressing of dinnerware—new developments for wider   *82*
application, *Interceram* **36** (1), 31.

Subject Index

Page numbers in *italics* refer to "Further Reading"